SEMICONDUCTOR LASER DIODE TECHNOLOGY AND APPLICATIONS

SEMICONDUCTOR LASER DIODE TECHNOLOGY AND APPLICATIONS

Editor

Brian Ismay

scitus
academics

Semiconductor Laser Diode Technology and Applications
Edited by **Brian Ismay**

ISBN: 978-1-68117-603-1
Library of Congress Control Number: 2016946458

616, Corporate Way, Suite 2, 4766,
Valley Cottage, NY 10989

www.scitusacademics.com

Printed in United States of America on Acid Free Paper

PREFACE

Lasers are in fact the subject of science fiction: large, heavy boxes that make blazing light blasts. If you've seen a regular laser in a laboratory before, you'll understand that it's a very heavy beast: usually as long as your forearms are hot, rather heavy and effective in producing a very intense light. But if the lasers are so large, why can we use them in small things such as portable CD players and handheld barcode scanners? A really different type of laser that is about the same size as (and works in a similar way to) a regular LED; these things are used as semiconductor lasers or possibly laser diode. A laser diode, or possibly LD, also known as an injection laser diode or possibly ILD, is an electrically pumped semiconductor laser in which the active laser medium is formed by a p n junction of a semiconductor diode similar to the one found in a light emitter. The laser diode is probably the most common type of laser produced for a wide range of purposes including fiber optic communications, barcode readers, laser pointers, reading and recording CD/DVD/Blue ray disks, laser printing, laser scanning and increasingly directional lighting sources. Laser diode semiconductor technology and applications are the latest developments in the rapidly evolving world of laser diode semiconductor technology and applications.

Semiconductor Laser Diode Technology and Applications represent the latest developments in the rapidly evolving world of laser diode semiconductor technology and applications. Lasers or laser diodes play an important role in our daily lives by providing low-cost and compact lasers. They consist of complex multi-layer structures that require the precision of the nanometer and an elaborate design. Their theoretical description is not only important from a fundamental point of view, but also to create new and improved designs. This book will be of great interest to engineers, scientists, technologists and physicists working in research and development in the field of laser semiconductor diode and to young researchers at the start of their careers.

TABLE OF CONTENTS

Chapter 1 Integrated Microwave Photonic Isolators: Theory, Experimental Realization and Application in a Unidirectional Ring Mode-Locked Laser Diode 1

Chapter 2 Bending effects of ZnO nanorod metal–semiconductor–metal photodetectors on flexible polyimide substrate 21

Chapter 3 Progress towards an Autonomous Field Deployable Diode-Laser-Based Differential Absorption Lidar (DIAL) for Profiling Water Vapor in the Lower Troposphere 39

Chapter 4 A high-performance quantum dot superluminescent diode with a two-section structure 69

Chapter 5 Frequency-converted dilute nitride laser diodes for mobile display applications 83

Chapter 6 A Novel Pulse Measurement System by Using Laser Triangulation and a CMOS Image Sensor 99

Chapter 7 Continuous in situ measurements of volcanic gases with a diode-laser-based spectrometer: CO_2 and H_2O concentration and soil degassing at Vulcano (Aeolian islands: Italy) 127

Chapter 8 Novel optoelectronic devices based on single semiconductor nanowires (nanobelts) 155

Chapter 9 **The use of laser in garment manufacturing: an overview** **175**

Chapter 10 **Diode Laser Assisted Filament Winding of Thermoplastic Matrix Composites** **211**

Chapter 11 **Steady-State Behavior of Semiconductor Laser Diodes Subject to Arbitrary Levels of External Optical Feedback** **225**

Chapter 12 **Diode Laser Detection of Greenhouse Gases in the Near-Infrared Region by Wavelength Modulation Spectroscopy: Pressure Dependence of the Detection Sensitivity** **245**

Index **267**

7

INTEGRATED MICROWAVE PHOTONIC ISOLATORS: THEORY, EXPERIMENTAL REALIZATION AND APPLICATION IN A UNIDIRECTIONAL RING MODE-LOCKED LASER DIODE

Martijn J.R. Heck [1,*], Sudharsanan Srinivasan [2], Michael L. Davenport [2] and John E. Bowers [2]

[1] Department of Engineering, Aarhus University, Aarhus 8200, Denmark

[2] Department of Electrical and Computer Engineering, University of California Santa Barbara, Santa Barbara 93106-9560, CA, USA

ABSTRACT

A novel integrated microwave photonic isolator is presented. It is based on the timed drive of a pair of optical modulators, which transmit a pulsed or oscillating optical signal with low loss, when driven in phase. A signal in the reverse propagation direction will find the modulators out of phase and, hence, will experience high loss. Optical and microwave isolation ratios were simulated to be in the range up to 10 dB and 20 dB, respectively, using parameters representative for the indium phosphide platform. The experimental realization of this device in the hybrid silicon platform showed microwave isolation in the 9 dB–22 dB range. Furthermore, we present a design study on the use of these isolators inside a ring mode-locked laser cavity. Simulations show that unidirectional operation can be achieved, with a 30–50-dB suppression of the counter propagating mode, at limited driving voltages. The potentially low noise and feedback-insensitive operation of such a laser makes it a very promising candidate for use as on-chip microwave or comb generators.

KEYWORDS

hybrid silicon platform; microwave photonics; mode-locked laser diodes; optical isolators; photonic integrated circuits; semiconductor optical amplifiers; silicon photonics

1. INTRODUCTION

The field of microwave photonics (MWP) utilizes photonics to generate, process, transport and detect radio frequencies (RF), typically in the gigahertz range [1]. This approach benefits from the large bandwidth and low loss signal propagation of photonic components, as compared to electronic components and circuits. Examples of successful MWP applications include high-bandwidth wireless links operating at 100 Gbps [2], low noise microwave oscillators [3], clock-free data generation [4] and photonic analog-to-digital converters for high bandwidth operation [5].

Such MWP systems are typically built using fiber-optic and discrete components, limiting their real-world applications, due to lack of robustness, high cost and large size and weight. Integration of MWP systems on a photonic integrated circuit (PIC) can in principle address these concerns and limitations. PIC technology has now evolved to the level that industry and academia have access to mature technology platforms [6]. This led to the field of integrated

MWP, where MWP functions are integrated on a PIC [7]. Recent advances, like the ultra-low waveguide losses, realized in the silicon nitride platform, can further unlock this potential, allowing for the almost fiber-like performance of a PIC-based system, e.g., for low noise microwave generation [8].

A main bottleneck for the integration of larger-scale systems on a PIC is the lack of practical integrated optical isolators, e.g., to prevent optical feedback from entering an on-chip laser cavity. For high-end MWP applications, low noise performance is of key importance, but such feedback results in increased phase and relative-intensity noise, increased optical linewidths or even chaotic laser emission [9,10]. Research efforts to realize an integrated isolator are mainly based on the integration of magneto-optic materials onto the PIC platform, to enable a non-reciprocal absorption or phase shift, however with a limited extinction ratio and high insertion loss [11]. Another way to break the symmetry between the forward and backward signals is by applying a time-dependent electro-optic index or transmission modulation, as presented in, e.g., [12,13]. More recently, isolation based on acousto-optic modulation was also shown, though without a clear path towards integration on mature PIC platforms [14].

The time-based asymmetry of a pulsed or MWP signal, propagating through a sequence of optical gates, which open and close periodically, can be used to achieve a difference in loss between forward and backward propagating signals. If a forward propagating oscillating signal passes through an open gate when its amplitude is maximum, its loss is minimal. However, if the spacing and timing between the gates is designed well, a backward propagating signal will not find all gates open and, hence, will incur additional losses. This general concept was theoretically studied in [15], where the self-amplitude modulation of a set of saturable absorption modulators (AMs) was used to create isolation between a pulsed signal and lower power feedback or noise.

In this work, we use a conceptually similar approach, but drive the modulators actively, by applying a time-varying voltage, effectively creating periodic gates. The concept and the simulation of the MWP isolator using realistic parameters, representative for indium phosphide-based active components, will be presented in Section 2. The realization in the hybrid silicon platform will be described in Section 3. Experimentally obtained results will be shown and discussed. In Section 4, we will theoretically investigate a possible promising implementation of the MWP in a ring mode-locked laser (MLL). It will be shown that unidirectional operation of the ring MLL can be achieved. To conclude, in Section 5, we will discuss the wider implementation and application range of these concepts.

2. MWP ISOLATOR CONCEPT AND THEORETICAL STUDY

The concept for an MWP isolator is shown in Figure 1, using two modulators. The modulators are driven with a period *T*, by varying the voltage over them, and, hence, have a time-dependent transmission. A forward propagating oscillating signal in phase with the transmission modulation of the modulators will experience minimum loss. This figure shows clearly how a backward propagating signal will always encounter at least one modulator in a (partially) blocking state, if the distance between the two modulators corresponds to $0.25 \cdot T \cdot v_g$, with v_g the group velocity of the signal in the system. This means that from the perspective of the backward propagating signal, the two modulator gates are always modulated out of phase with each other, whereas from the perspective of the forward propagating signal, these modulator gates are modulated in phase.

Although this concept would work with any type of RF-driven modulator, we limit the discussion to reversely-biased semiconductor optical amplifier (SOA) sections, effectively creating AMs, as was done in [12]. The main reason is the obvious compatibility with the amplifying SOA sections in a PIC, which allows for integration of the MWP isolator after the output of a laser and even into a laser cavity, as will be shown in Section 4. As mentioned above, this is the main rationale for an integrated isolator. From a fabrication perspective, this means that no epitaxial regrowth or overgrowth steps are required, nor techniques like quantum-well intermixing. Experimentally, it was validated that the SOA transmission, more specifically the absorption, changes with applied reverse bias voltage, and hence, it is feasible to use it as an AM [16]. Note that in this implementation, the operating wavelength is well below the bandgap wavelength of the AM, which is different from what is typically called an electro-absorption modulator (EAM), in which case the wavelength is close to the bandgap.

We can model the MWP isolator, as shown in Figure 1, using SOA rate equations for the optical field $A(z,t)$ and the carrier density $N(z,t)$ inside the SOA as presented in, e.g., [12,17]:

$$\frac{\partial A(z,t)}{\partial z} \pm \frac{1}{v_g}\frac{\partial A(z,t)}{\partial t} = \frac{1}{2}\frac{G(z,t)}{1+\varepsilon P(z,t)}A(z,t) - \frac{i}{2}\alpha_N G(z,t)A(z,t) - \frac{1}{2}\alpha_{int}A(z,t) + F_{SE}(z,t) \quad (1)$$

$$\frac{\partial N(z,t)}{\partial t} = \frac{I}{qV} - \frac{N(z,t)}{\tau} - \frac{1}{\Gamma\hbar\omega\sigma}\frac{G(z,t)P(z,t)}{1+\varepsilon P(z,t)} \quad (2)$$

with the gain $G(z,t)$ expressed as a linearized dependency on the carrier density $N(z,t)$:

$$G(z,t) = \Gamma a_N\left(N(z,t) - N_{tr}\right) \quad (3)$$

Here, the optical field $A(z,t)$ is expressed in [$W^{1/2}$], $P(z,t)$ is the optical power in [W], ε is the nonlinear gain compression (0.2 W^{-1} for SOA, 0 for AM, *i.e.*, reverse bias), α_N is the linewidth enhancement factor (3 for SOA, 1 for AM), α_{int} = 20 $cm^{-1}$the passive loss, $F_{SE}(z,t)$ a Langevin noise term representing spontaneous emission, I the injection current, q the elementary charge, V the volume of the active area, τ the carrier recovery time (300 ps for SOA, 5 ps for AM), Γ = 0.3 the confinement factor, $\hbar$ is Planck's constant, ω the optical angular frequency, σ the mode cross section, a_N the differential gain coefficient (3×10^{-20} m^2 for SOA, 12×10^{-20} m^2 for AM) and N_{tr} the transparency carrier density (0.3×10^{18} cm^{-3} for SOA, 0.7×10^{18} cm^{-3}for AM). These parameters are representative for the indium phosphide technology and have been experimentally verified, e.g., [13]. The absorption dependency of the AM on the bias voltage is modeled using a voltage-dependent transparency carrier density $N_{tr} = N_{tr}(V(t))$, as proposed in [18] and experimentally verified in [13]. These equations are numerically solved using a spatial resolution of $\Delta z = 5$ μm, corresponding to a time resolution of $\Delta t = 60$ fs.

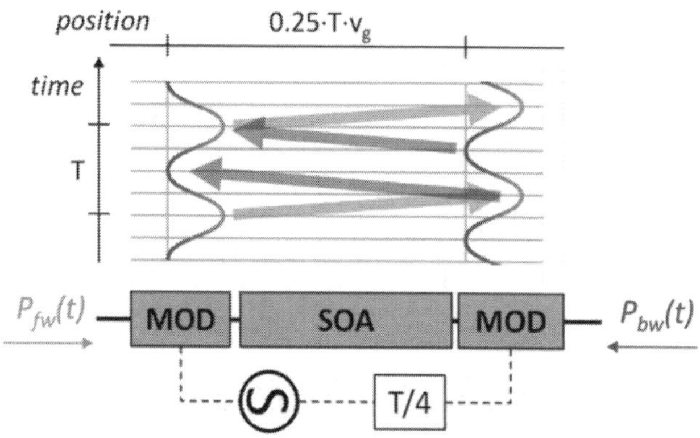

Figure 1. Schematic of the microwave photonics (MWP) isolator concept using modulators (MODs), showing the concatenated MOD-semiconductor optical amplifier (SOA)-MOD array, with the forward propagating signal P_{fw} entering from the left and the backward propagating signal P_{bw} entering from the right. The MOD gates are driven with a $T/4$ time difference, with T the period of the RF signal. The MOD transmission is indicated schematically with the sinusoid, with period T, with red indicating low transmission and green indicating high transmission. The green and red arrows are following the time-position grid and show how a forward propagating signal (green arrow) can pass two open gates, whereas a backward propagating signal (red arrow) always encounters at least on (partially) closed gate. The optical amplifier (SOA) can be used to compensate for overall insertion losses.

Using this model, the feasibility of the MWP isolator can be explored. The transmission of the forward and backward propagating 20-GHz MWP fields for an isolator with 100 μm-long AMs is shown in Figure 2a. Obviously, the AM driving voltage needs to be in phase with the MWP signal oscillator for maximum transmission, as shown by the maximum around three radians drive offset. Note that this is a matter of calibrating the model, or any real-world system, and represents a relative value only. Since a system integrated on a PIC has known path lengths, the required drive offset is known beforehand. On the other hand, the counter-propagating signal has in practice an unknown phase with respect to the AMs, as reflections can originate from anywhere downstream of the circuit. However, as can be seen in Figure 2a, this transmission is flat and below the peak transmission of the forward signal, indicating the isolation capability. Note that this flat transmission also means that isolation from continuous-wave counter-propagating signals is achieved.

In Figure 2b, the isolation is plotted for various AM lengths and various driving voltage amplitudes of 3–9 V (corresponding to N_{tr} = (0.1–0.3) × 10^{18} cm^{-3}; [13]). As can be expected, isolation increases with AM length and driving voltage amplitude. However, it has to be noted that such driving voltages are quite high, requiring between 100 mW and 1 W of RF power on the AM, assuming a 50-Ω termination. Based on these results, it can be seen that modest optical isolation ratios up to 10 dB can be achieved, which corresponds to close to a 20-dB microwave isolation, as most of the power is in the 20-GHz microwave fundamental harmonic, with up to 20% in the 40-GHz harmonic, due to the nonlinear transmission.

3. EXPERIMENTAL REALIZATION IN HYBRID SILICON TECHNOLOGY

The MWP isolator was fabricated in the hybrid silicon platform [19]. In this platform, an indium phosphide layer is bonded to a processed silicon photonic chip and then processed further, ensuring lithographic alignment accuracy. Silicon photonic components, such as waveguides, splitters and multiplexers, can be combined in this platform with hybrid waveguides, where the optical mode overlaps with the indium phosphide layers. In this way, active optical functionalities, such as gain, can be achieved. A schematic of the device cross-section is shown in Figure 3a, where it can be seen that the optical mode propagates through an AM-SOA-AM sequence. The tapered mode converters from the silicon waveguide to the hybrid waveguide are electrically pumped, hence the small SOA sections at the start and end of the device [20]. For details about the fabrication process, we refer to [16]. The chip is diced, polished and anti-reflection coated for characterization.

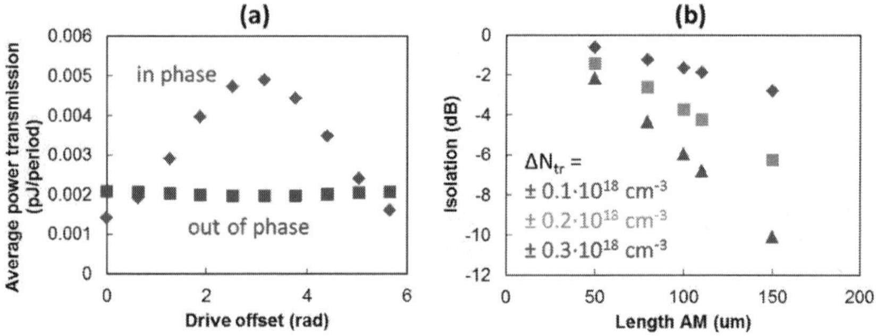

Figure 2. (**a**) Power transmission of a 1-mW peak power 20-GHz MWP signal through an isolator with a total length of 1.2 mm and with two absorption modulator (AM) sections of 100 μm. The bias amplitude corresponds to ±0.2 × 10^{18} cm^{-3}. Referring toFigure 1, the forward transmission (red) and reverse, backward transmission (blue) are shown, as a function of the phase difference between the MWP signal and the AM driving voltage. (**b**) Maximum difference between forward and backward propagating signals for various AM lengths and driving voltages (indicated with N_{tr}).

The experimental setup to characterize the MWP isolator is shown in Figure 3c. An MWP signal is generated by an oscillator that is synchronized with the AM drivers. Two microwave phase shifters allow for independent tuning of the phase of the driving voltage on the AMs, compared to the MWP signal. The optical carrier wavelength is 1583 nm, as the AM has maximum dynamic range here, expressed as absorption change as a function of applied voltage. This was determined by looking at the amplified spontaneous emission spectrum of the SOA through the AM, on which the bias was varied, as shown in Figure 4a. The MWP isolator insertion loss is 4.4 ± 0.5 dB, which of course depends on the SOA bias, in this case 200 mA. The overall active device length is 1.4 mm, with two 100-μm long AMs, positioned with a 1.0-mm spacing.

Figure 4b shows the 20-GHz MWP signal transmission with $\Delta\varphi_1$ set to drive one AM in phase with the MWP signal, while tuning the phase of the voltage on the other AM by $\Delta\varphi_2$ (Figure 3b). As can be seen, the received RF power varies with changing the phase on the second AM. Using this plot as a calibration, it can be seen that at $\Delta\varphi_2 = 50° \pm 10°$, the two AMs are driven at $\Delta T = T/4$, as explained in Figure 1, with $T = 50$ ps. This means maximum transmission for the forward signal and minimum transmission for the counter-propagating signal. At $\Delta\varphi_2 = 230° \pm 10°$, this situation is reversed, as expected, and represents $\Delta T = -T/4$. From this plot, it can be seen that 22-dB and 9-dB microwave isolations have been achieved, which agrees well with the optical isolation ratios up to 10 dB, obtained using our simulation. This comparison should be considered qualitative, as a complete set of SOA and AM parameters for the hybrid silicon platform is not available, and the simulation used parameters from the monolithic indium phosphide technology. However, since the active part of the hybrid waveguide is indium phosphide and since the simulation tool uses parameter values experimentally obtained with an indium phosphide SOA, it can be assumed that these outcomes should be close.

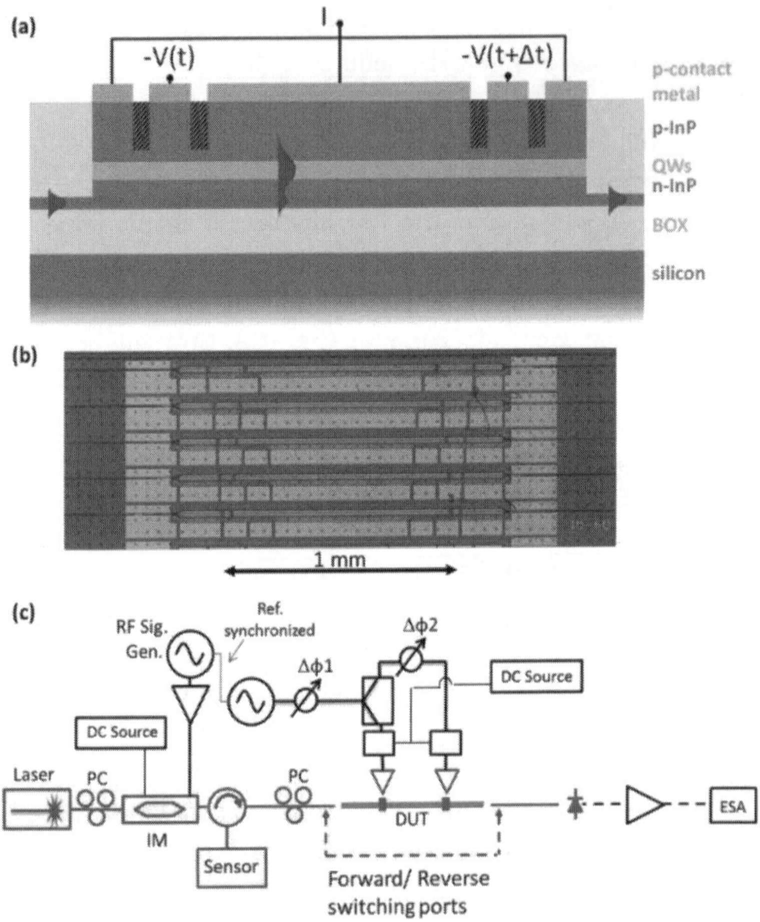

Figure 3. (**a**) Schematic of the hybrid silicon MWP isolator, showing the optical field (red) as it propagates through the silicon waveguide, hybrid silicon III/V waveguide and the silicon waveguide again. Hydrogen-implantation of the p-doped InP cladding provides electrical isolation between the forward-biased SOA and the reverse-biased AM. N-contacts are to the side and not shown in this cross-section. (**b**) Microscope picture of an array of realized devices. (**c**) Experimental setup, showing the MWP signal generation using an intensity modulator (IM, $V_\pi = 3.6$ V), driven by a signal generator (RF Sig. Gen.). This signal generator is synchronized with another signal generator that drives the two AMs. Two microwave phase shifters ($\Delta\varphi_{1,2}$) are used to control the phase. The signal is recorded using an

electrical spectrum analyzer (ESA) connected to a 40-GHz bandwidth photodiode and a 20-dB RF amplifier. Polarization controllers (PCs) are used to ensure transverse-electric polarization of the field inside the device under test (DUT). Lensed fibers are used to couple the light in and out of the device. Electrical connections are shown in black, optical connections in red. An estimated power of 12.5 dBm is used to drive each of the AMs. The SOA is biased at 200 mA. The circulator with the sensor is used for fiber to chip alignment.

We note that ideally, the operation of the device should be symmetric for forward and reverse propagation. From Figure 4b, it can be seen that this is not the case in the experiments. We hypothesize that due to processing non-uniformities, most notably variations in contact resistance over the length of the device, the AMs do not operate identically. In the linear regime, that would still not matter, but as the AM and SOA are nonlinear devices, *i.e.*, their transmission is power dependent, a difference between the two AMs would lead to a difference between the forward and reverse signals, even when both AMs are driven in phase.

4. APPLICATION OF THE MWP ISOLATOR IN UNIDIRECTIONAL RING MODE-LOCKED LASERS

The theoretical and experimental results show clearly the potential for isolation, though with a somewhat limited extinction ratio up to ~10 dB optical and ~20 dB electrical. One possible application could be to use this MWP isolator inside a ring laser cavity. It is well known that a small periodic variation in the gain can cause mode-locking, *i.e.*, the laser operates in a pulsed mode [21]. Such MLLs have been realized in various configurations, but ring lasers are particularly attractive due to the lack of mirrors and, hence, the ease of integration on a PIC. However, such ring MLLs are intrinsically bi-directional [18], which leads to lower output efficiency and higher amplitude and phase noise due to cross-gain modulation. It has been shown theoretically and experimentally that by careful MLL cavity design, the power ratio in two counter propagating fields can be controlled, however only to a limited extent [22,23].

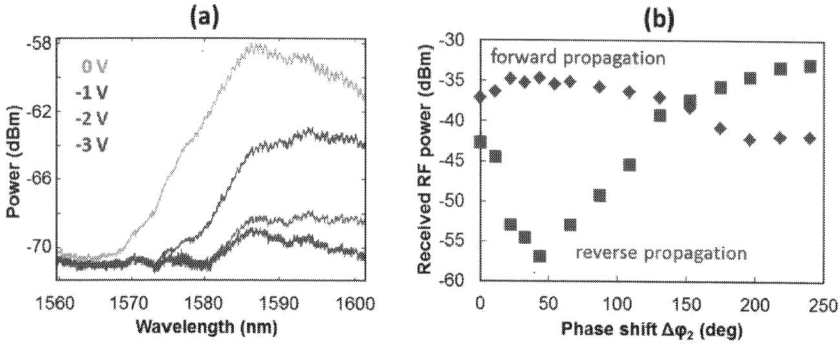

Figure 4. (**a**) Amplified spontaneous emission spectra, obtained by biasing the SOA with a constant current and measured after propagation through an AM. Bias voltage on the AM is varied from 0–3 V. (**b**) The RF power measured with the electrical spectrum analyzer for the forward and backward propagating MWP signals as a function of the phase difference between the two AM drive voltages.

In this section, we explore theoretically the feasibility of using the MWP isolator to force the MLL into a unidirectional regime. We use the model presented above (Equations (1)–(3)), extended with a 5-THz SOA gain bandwidth filter, to limit the pulse shortening. The model is implemented using a similar approach as in [19]. Figure 5a shows the layout of a ring MLL with a dual AM-phased drive, which means that the MLL is effectively hybridly mode-locked, as the AMs are saturable absorbers and this saturation contributes to the pulse shaping mechanism. The length of the ring determines the period T of the output pulses, and the AMs are placed with a separation corresponding to $T/4$. Figure 5b,c shows the simulated clockwise (CW) and counter-clockwise (CCW) optical output fields for a 20-

GHz ring MLL and with an AM drive delay of 12.5 ps and −12.5 ps, respectively. As can be seen, the ring MLL operates unidirectionally with the correct phase of the drive voltage on the AMs. The isolation or suppression of the counter propagating field depends on the MLL operating conditions. In Figure 6, this is shown for a MLL with varying drive voltage amplitude, here again expressed by varying the amplitude of the oscillation of the transparency carrier density N_{tr}. The suppression of the counter propagating field ranges from 30 dB–50 dB, for increasing drive voltage amplitude.

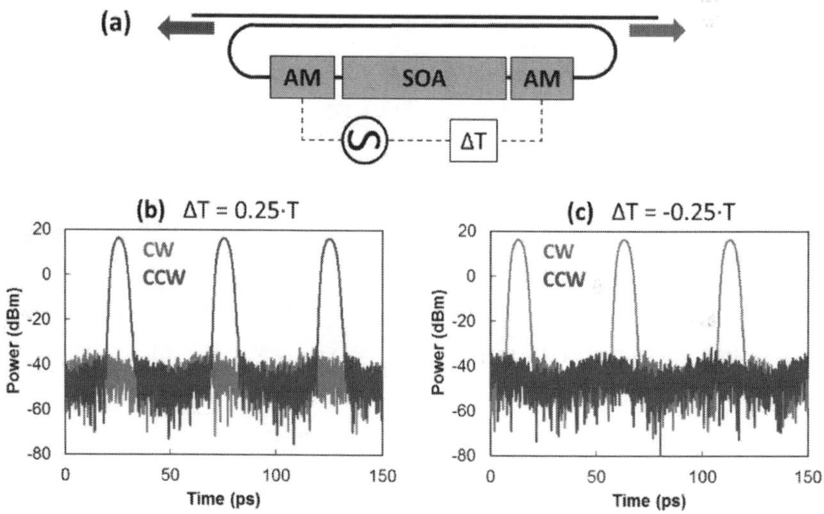

Figure 5. (**a**) Schematic layout of the unidirectional mode-locked laser (MLL), showing an oscillator driving the two AMs, with a relative delay of ΔT. A directional coupler couples light out of the ring into the clockwise (CW, blue) or counterclockwise (CCW, red) output waveguide. The simulated output of a 20-GHz MLL is shown for (**b**) ΔT = 12.5 ps and (**c**) ΔT = −12.5 ps. The SOA length is 800 μm, and the AMs have a length of 25 μm. Power output coupling is 0.5 per roundtrip. The simulation parameters are ΔN_{tr} = 0.7 × 10^{18} cm^{-3} and I_{SOA} = 80 mA.

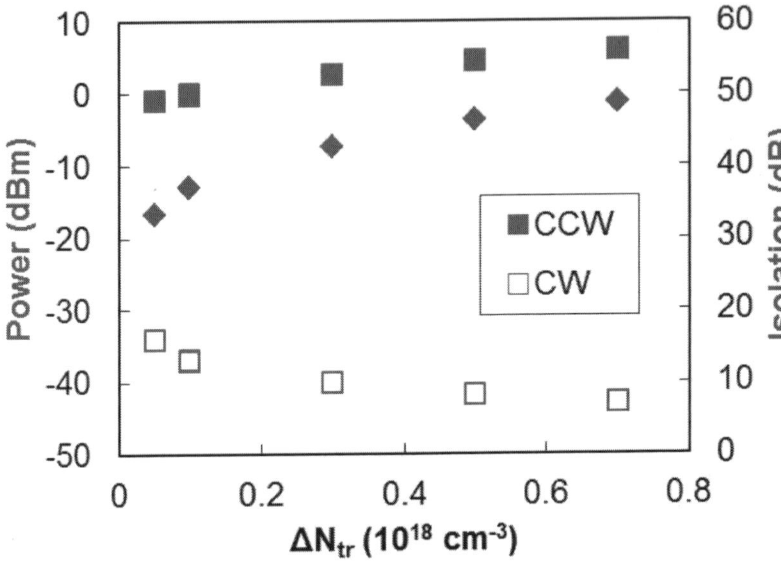

Figure 6. Average output power for the clockwise (CW) and counter-clockwise (CCW) signals, assuming ΔT = 12.5 ps in Figure 5a, as a function of the modulation amplitude ΔN_{tr} of N_{tr}. The isolation ratio is also shown (blue). The SOA injection current is 80 mA. Further MLL design parameters are as in Figure 5.

In conclusion, we have shown the feasibility of an integrated unidirectional ring MLL, using our experimentally-verified concept of an MWP isolator. Apart from a more efficient output, the lack of cross-gain modulation and spatial hole burning will likely lead to lower phase and amplitude noise. Furthermore, it can be hypothesized that the laser is less sensitive to optical feedback, as this feedback is injected in the suppressed counter propagating mode. Major applications can be foreseen in on-chip low noise microwave generation and in on-chip comb generation for wavelength-division multiplexed systems.

5. CONCLUSION AND OUTLOOK

In this paper, we have presented the concept of a novel integrated MWP isolator and its first experimental realization in the hybrid silicon platform. Optical and microwave isolation ratios were simulated to be in the range up to 10 dB and 20 dB, respectively, using parameters representative for the indium phosphide platform. The experimental realization of this device in the hybrid silicon platform showed microwave isolation in the 9 dB–22 dB range.

Such isolation ratios are barely sufficient for on-chip use, for example to shield a laser from optical feedback. As was argued in [24], optical isolation of at least 20 dB is required. Options to enhance the MWP isolator include the concatenation of more AM gates and using more square-shaped driving pulses, which would also reduce higher harmonic generation.

In an alternative implementation, when using this MWP isolator inside a resonant cavity, the isolating effect is strongly enhanced. We have presented a design study on the use of MWP isolators inside a ring MLL cavity. Simulations show that unidirectional operation can be achieved, with a 30–50-dB suppression of the counter propagating mode, at limited driving voltages. The potentially low noise and feedback-insensitive operation of such a laser makes it a very promising candidate for use as on-chip microwave or comb generators.

The simulations were based on the indium phosphide material system, and an implementation using reversely-biased SOA sections or AMs was chosen. This choice offers a practical path to integration with (forward biased) SOAs, for example to realize the unidirectional ring MLLs that were presented in this work. Alternative approaches for the AMs can include Mach–Zehnder interferometer modulators in the silicon photonics platform, electro-absorption modulators in the indium phosphide and hybrid silicon platforms or even forward-biased SOAs as gates. The choice will depend on all trade-offs that are required for PIC design and fabrication. Choosing modulators with a higher extinction ratio will improve the isolation ratio. Another important aspect is the optical bandwidth that is required. Mach–Zehnder interferometer modulators and forward-based SOAs are broadband, having bandwidths of at least tens of nanometers. Electro-absorption modulators and reversely-biased SOAs have a very wavelength-dependent transmission and are hence narrow band, typically only a few nanometer, e.g., as shown in [25] and in Figure 4.

The experimental realization on the hybrid silicon platform is a relevant step, as this platform has been shown to have interesting applications in the field of MWP [26]. Although one can argue that similar PIC functionalities can be achieved using a monolithic indium phosphide platform, the hybrid silicon platform offers the potential of economies of scale, in terms of fabrication on 6" or even 8" substrate wafers and the compatibility with the mature CMOS manufacturing infrastructure. This also includes the required integration with electronic driving and control circuitry, which can be done in a package [27] or fully monolithically on the same chip [28]. Such an approach is necessary for large volume applications that can be envisioned for MWP, which include, for example, 5G wireless communications infrastructure.

ACKNOWLEDGMENTS

The authors would like to thank Greg Fish and Erik Norberg from Aurrion Inc. for helpful discussions. This work was supported by the DARPA EPHI Project. M.J.R.H. acknowledges partial funding from Det Frie Forskningsråd through the mmW-SPRAWL project (DFF-4005-00246).

AUTHOR CONTRIBUTIONS

M.J.R.H. conceived of the idea and did the simulations and mask design. M.L.D. developed the fabrication process and fabricated the devices. S.S. performed the experimental characterization. M.J.R.H. and J.E.B. supervised the project. M.J.R.H. wrote the paper, with input from all other co-authors.

CONFLICT OF INTEREST

The authors declare no conflict of interest.

REFERENCES

1. Yao, J. Microwave photonics. *J. Lightw. Technol.* **2009**, *3*, 314–335.

2. Pang, X.; Caballero, A.; Dogadaev, A.; Arlunno, V.; Borkowski, R.; Pedersen, J.S.; Deng, L.; Karinou, F.; Roubeau, F.; Zibar, D.; *et al.* 100 Gbit/s hybrid optical fiber-wireless link in the W-band (75–110 GHz). *Opt. Express* **2011**, *19*, 24944–24949.

3. Yao, X.S.; Maleki, L. Optoelectronic microwave oscillator. *JOSA B* **1996**, *13*, 1725–1735.

4. Chi, Y.-C.; Lin, G.-R. A self-started laser diode pulsation based synthesizer-free optical return-to-zero on-off-keying data generator. *IEEE Trans. Microwave Theory Tech.* **2010**, *58*, 2292–2298.

5. Ghelfi, P.; Laghezza, F.; Scotti, F.; Serafino, G.; Capria, A.; Pinna, S.; Onori, D.; Porzi, C.; Scaffardi, M.; Malacarne, A.;*et al.* A fully

photonics-based coherent radar system. *Nature* **2014**, *507*, 341–345.

6. Smit, M.K.; Leijtens, X.; Bente, E.; Van der Tol, J.; Ambrosius, H.; Robbins, D.; Wale, M.J.; Grote, N.; Schell, M. A generic foundry model for InP-based photonic ICs. In Proceedings Optical Fiber Communication Conference OM3E-3, Los Angeles, LA, USA, 4–8 March 2012.

7. Marpaung, D.; Roeloffzen, C.; Heideman, R.; Leinse, A.; Sales, S.; Capmany, J. Integrated microwave photonics. *Laser Photonics Rev.* **2013**, *7*, 506–538.

8. Heck, M.J.R.; Bauters, J.F.; Davenport, M.L.; Spencer, D.T.; Bowers, J.E. Ultra-low loss waveguide platform and its integration with silicon photonics. *Laser Photonics Rev.* **2014**, *8*, 667–686.

9. Tkach, R.W.; Chraplyvy, A.R. Regimes of feedback effects in 1.5 μm distributed feedback lasers. *J. Lightw. Technol.* **1986**, *4*, 1655–1661.

10. Barbarin, Y.; Bente, E.A.J.M.; Heck, M.J.R.; Den Besten, J.H.; Guidi, G.; Oei, Y.S.; Binsma, J.J.M.; Smit, M.K. Realization and modeling of a 27-GHz integrated passively mode-locked ring laser. *IEEE Photonics Technol. Lett.* **2005**, *17*, 2277–2279.

11. Mizumoto, T.; Takei, R.; Shoji, Y. Waveguide optical isolators for integrated optics. *IEEE J. Quant. Elect.* **2012**, *48*, 252–260.

12. Galland, C.; Ding, R.; Harris, N.C.; Baehr-Jones, T.; Hochberg, M. Broadband on-chip optical non-reciprocity using phase modulators. *Optics Express* **2013**, *21*, 14500–14511.

13. Tzuang, L.D.; Fang, K.; Nussenzveig, P.; Fan, S.; Lipson, M. Non-reciprocal phase shift induced by an effective magnetic flux for light. *Nat. Photonics* **2014**, *8*, 701–705.

14. Li, E.; Eggleton, B.J.; Fang, K.; Fan, S. Photonic Aharonov-Bohm effect in photon-phonon interactions. *Nat. Commun.* **2014**, *5*, 1–5.

15. Heck, M.J.R.; Bente, E.A.J.M.; Barbarin, Y.; Lenstra, D.; Smit, M.K. Monolithic Semiconductor Waveguide Device Concept for Picosecond Pulse Amplification, Isolation and Spectral Shaping. *IEEE J. Quant. Elect.* **2007**, *43*, 910–922.

16. Heck, M.J.R.; Bente, E.A.J.M.; Barbarin, Y.; Fryda, A.; Jung, H.-D.; Oei, Y.S.; Nötzel, R.; Lenstra, D.; Smit, M.K. Characterization of a monolithic concatenated SOA/SA waveguide device for picosecond pulse amplification and shaping. *IEEE J. Quant. Elect.* **2008**, *44*, 360–369.

17. Jones, D.J.; Zhang, L.M.; Carroll, J.E.; Marcenac, D.D. Dynamics of monolithic passively mode-locked semiconductor lasers. *IEEE J. Quant. Elect.* **1996**, *31*, 1051–1058.

18. Bandelow, U.; Radziunas, M.; Vladimirov, A.; Hüttl, B.; Kaiser, R. 40 GHz mode-locked semiconductor lasers: Theory, simulations and experiment. *Opt. Quant. Elect.* **2006**, *38*, 495–512.

19. Heck, M.J.R.; Bauters, J.F.; Davenport, M.L.; Doylend, J.K.; Jain, S.; Kurczveil, G.; Srinivasan, S.; Tang, Y.; Bowers, J.E. Hybrid silicon photonic integrated circuit technology. *IEEE J. Sel. Top. Quant. Elect.* **2013**, *19*.

20. Kurczveil, G.; Pintus, P.; Heck, M.J.R.; Peters, J.D.; Bowers, J.E. Characterization of insertion loss and back reflection in passive hybrid silicon tapers. *IEEE Photonics J.* **2013**, *5*.

21. Barbarin, Y.; Bente, E.A.J.M.; Heck, M.J.R.; Oei, Y.S.; Nötzel, R.; Smit, M.K. Characterization of a 15 GHz integrated bulk InGaAsP passively modelocked ring laser at 1.53 μm. *Opt. Express* **2006**, *14*, 9716–9727.

22. Bente, E.A.J.M.; Barbarin, Y.; Heck, M.J.R.; Smit, M.K. Modeling of integrated extended cavity InP/InGaAsP semiconductor modelocked ring lasers. *Opt. Quant. Elect.* **2008**, *40*, 131–148.

23. Tahvili, M.S.; Barbarin, Y.; Leijtens, X.J.M.; De Vries, T.; Smalbrugge, E.; Bolk, J.; Ambrosius, H.P.M.M.; Smit, M.K.; Bente, E.A.J.M. Directional control of optical power in integrated InP/InGaAsP extended cavity mode-locked ring lasers. *Opt. Lett.* **2011**, *36*, 2462–2464.

24. Heck, M.J.R. Grating Coupler Enabled Optical Isolators and Circulators for Photonic Integrated Circuits. *IEEE J. Sel. Top. Quant. Elect.* **2015**, *21*, 1–9.

25. Tang, Y.; Chen, H.-W.; Jain, S.; Peters, J.D.; Westergren, U.; Bowers, J.E. 50 Gb/s hybrid silicon traveling-wave electroabsorption modulator. *Opt. Express* **2011**, *19*, 5811–5816.

26. Fish, G. Heterogeneous photonic integration for microwave photonic applications. In Proceedings Optical Fiber Communication Conference and Exposition and the National Fiber Optic Engineers Conference (OFC/NFOEC), Anaheim, CA, USA, 17–21 March 2013.

27. Dupuis, N.; Lee, B.G.; Proesel, J.E.; Rylyakov, A.; Rimolo-Donadio, R.; Baks, C.W.; Ardey, A.; Schow, C.L.; Ramaswamy, A.; Roth, J.E.; *et al.* 30-Gb/s Optical Link Combining Heterogeneously Integrated III-V/Si Photonics With 32-nm CMOS Circuits. *J. Lightw. Technol.* **2015**, *33*, 657–662.

28. Thomson, D.J.; Porte, H.; Goll, B.; Knoll, D.; Lischke, S.; Gardes, F.Y.; Hu, Y.; Reed, G.T.; Zimmermann, H.; Zimmermann, L. Silicon carrier depletion modulator with 10 Gbit/s driver realized in high-performance photonic BiCMOS. *Laser Photonics Rev.* **2014**, *8*, 180–187.

2

BENDING EFFECTS OF ZNO NANOROD METAL–SEMICONDUCTOR–METAL PHOTODETECTORS ON FLEXIBLE POLYIMIDE SUBSTRATE

Tse-Pu Chen, Sheng-Joue Young, Shoou-Jinn Chan, Chih-Hung Hsiao and Yu-Jung Hsu

1 Institute of Microelectronics and Department of Electrical Engineering, Center for Micro/Nano Science and Technology, Advanced Optoelectronic Technology Center, National Cheng Kung University, Tainan, 701, Taiwan.

2 Department of Electronic Engineering, National Formosa University, Yunlin, 632, Taiwan. 3Institute of Electro-Optical Science and Engineering, National Cheng Kung University, Tainan, 701, Taiwan.

ABSTRACT

The authors report the fabrication and I-V characteristics of ZnO nanorod metal–semiconductor–metal photodetectors on flexible polyimide substrate. From field-emission scanning electron microscopy and X-ray diffraction spectrum, ZnO nanorods had a (0002) crystal orientation and a wurtzite hexagonal structure. During the I-V and response measurement, the flexible substrates were measured with (i.e., the radius of curvatures was 0.2 cm) and without bending. From I-V results, the dark current decreased, and the UV-to-visible rejection ratio increased slightly in bending situation. The decreasing tendency of the dark current under bending condition may be attributed to the increase of the Schottky barrier height.

KEYWORDS

ZnO, Nanorod, MSM, Photodetector, Flexible

1. BACKGROUND

Zinc oxide (ZnO), a nanostructured material that has been widely investigated, has a wide energy band gap of 3.37 eV at room temperature, high optical gain of 300/cm which is higher than that of GaN (100/cm) [1], and large exciton binding energy of 60 meV [2] which is higher than that of ZnSe (22 meV) and GaN (25 meV). The large exciton binding energy provides high-luminescence efficiency of light emission at or above room temperature. ZnO has slightly higher saturation velocity of 3.2×10^7 cm/s [3] than GaN, InGaN, and AlGaN [4, 5], but the room temperature electron Hall mobility (205 cm^2/V/s/) [6] is lower than that of GaN. ZnO has high mechanical and thermal stabilities, and radiation hardness for devices used in nuclear and space applications. ZnO also has lower growth temperature and material cost than III-nitride materials.

According to the researches, GaN and its alloys with AlN and InN cover the spectral range from red to vacuum UV (1.9 to 6.2 eV). Therefore, III-nitride materials have attracted a great deal of attention since the commercialization of light-emitting diodes. Not only GaN but also ZnO can tune the value of band gap by forming the ternary alloy of ZnMgO and ZnCdO with MgO and CdO [7], respectively.

Over the past decade, ZnO-based and III-nitride-based light-emitting diodes (LEDs) and laser diodes (LDs) have attracted much interest for display, illumination, and mobile phone backlights. Recently, Zhang et al. [8] studied on gain properties of high Al-content AlGaN-delta-GaN quantum wells (QWs) for mid- and deep-UV lasers. Zhao et al. [9] investigated the QW structures with large overlap design to enhance the internal quantum efficiency for InGaN QW-based LEDs. Shukla [10] studied a p-n junction LED employing ZnO/MgZnO QW active layer on a *c*-plane sapphire by the pulsed-laser deposition technique. Ahn et al. [11] showed p-n heterojunction LEDs that were formed from a p-Si thin film/nanostructured n-ZnO by a dielectrophoresis method. Except for the LEDs and LDs, ZnO makes it as a promising functional material for the electronic device manufacture such as field emission, photodetectors (PD), solar cells, waveguides, and chemical or biosensors.

To date, many groups have reported encouraging results for ZnO-based photodetectors. It is important that ZnO-based PDs be used in various military and commercial applications, like missile launching and flame detections, optical communications, and ozone layer monitoring. Indeed, ZnO-based PDs have various types, such as p-n junction PDs, p-i-n PDs, Schottky barrier PDs, and metal–semiconductor–metal (MSM) PDs [12–15]. Among the above structures, the MSM structure is a practical application due to its easy fabrication, low dark current and device noise values, high-response speed, and compatibility with integrated circuit technology. Recently, Yen et al. [16] obtained high photocurrent generation with ZnO/Si heterostructure MSM PDs by an avalanche multiplication in the ZnO layer. GaN-based MSM PDs also can detect the UV region, but a significant number of threading dislocations exit in GaN epilayers due to the large mismatches in lattice constant and thermal expansion coefficient between GaN and substrate, like sapphire and Si. Recently, Li et al. [17] found that dislocations had strong influence on the dark current and responsivity of the PDs. Consequently, ZnO-based materials have more advantages for PDs than GaN-based materials.

Recently, various one-dimensional (1-D) ZnO semiconducting nanostructures have been synthesized like nanorods, nanotubes, and nanobelts [18–20]. Most importantly, ZnO is well known as a piezoelectric material. Because of semiconductor properties and the coupling of piezoelectric of ZnO, many groups extensively investigated nanogenerators, piezoelectric field effect transistors, and piezoelectric strain and piezoelectric humidity/chemical sensors [21–24].

Today, many groups investigate the optoelectronic devices grown on flexible substrates, but the problem of these flexible substrates is that they must use low-temperature process. ZnO is one of the most suitable materials on flexible substrate because it can be prepared by low temperature processes, like sputter and aqueous method. Growth of 1-D ZnO semiconducting nanostructures on flexible substrates has been widely reported which are ideal candidates for LEDs, solar cell, file-effect transistors, personal health monitors, and field emission [22, 25–27] due to cheap, lightweight, and portable characteristics. However, the mechanical bending characteristics of the ZnO nanorod MSM structures fabricated on flexible substrates have rarely been investigated.

In this letter, we fabricated MSM ultraviolet PDs with ZnO nanorods on flexible substrate by aqueous method. The advantages of aqueous method are low cost, simple process, low temperature, and high product yield. Further, the fabricated devices were measured with and without strong mechanical bending. The optical and electric properties and the working principle of the ZnO nanorod MSM PDs would be presented in detail.

2. METHODS

The polyimide (PI) substrates were ultrasonically cleaned in methanol, isopropanol, and deionized water and dried with nitrogen. ZnO-seed layers with a thickness of about 130 nm were grown on polyimide substrates in a low-pressure RF sputter reactor. In our sputtering system, the background pressure in the sputtering chamber was evacuated to about 8×10^{-6} Torr with a turbo pump. The deposition processes were performed in Ar and O_2 ambient at an applied RF power of 120 W. After deposition, the as-grown samples were annealed 40 min in oxygen flow at 300°C in a quartz tube. After annealing, the samples were cooled to room temperature naturally, and MSM PDs were subsequently fabricated on the annealed samples. The ZnO PDs were fabricated using interdigital MSM structures by standard lithography. Using e-beam evaporation, Ag/Au (40/50 nm) contact electrodes were deposited on the samples served as Schottky contacts. The metal contact did not anneal by furnace or RTA. Then, we used photoresist to protect the electrodes by standard lithography. ZnO nanorod arrays were grown using an aqueous solution containing zinc nitrate ($Zn(NO_3)_2$) and hexamethylenetetramine (HMTA) for 1 h at 90°C. During ZnO nanorod synthesis, $Zn(NO_3)_2$and HMTA act as a Zn^{2+} source and a pH buffer, respectively, which keep the pH value at a constant [28]. After the process of the grown ZnO nanorods, the photoresist was lift-off, and there is no ZnO film or short nanorod formed on the metal electrode.

Figure 1 shows the schematic of the proposed ZnO nanorod MSM photodetector. The morphologies and size distribution of the ZnO nanorod PDs were characterized by field-emission scanning electron microscopy (JSM-7000 F, JEOL Ltd., Akishima, Tokyo, Japan) which was operated at 10 keV. X-ray diffraction (XRD) measurement was then utilized to characterize the optical and crystallographic properties of the as-grown ZnO nanorods. An HP-4156 C semiconductor parameter analyzer (Agilent Technologies Inc., Santa Clara, CA, USA) was then employed to measure current–voltage (I-V) characteristics of the proposed ZnO nanorod MSM photodetector. The photo-responsivity was obtained using the TRIAX 180 system (HORIBA Ltd., Minami-Ku, Kyoto, Japan) with a 300 W xenon arc lamp light source and standard synchronous detection scheme.

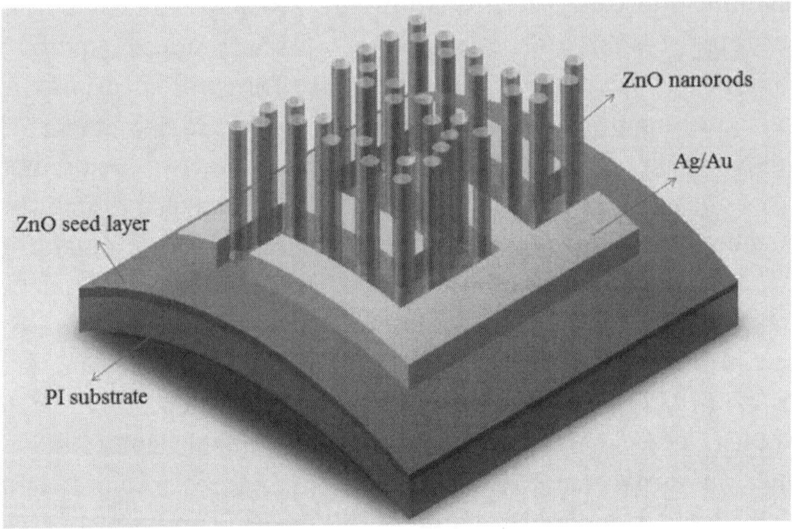

Figure 1 Schematic of the ZnO nanorod MSM PD prepared on flexible substrate.

3. RESULTS AND DISCUSSION

Figure 2a and its inset shows the top view and cross-sectional SEM images of the ZnO nanorods grown on the flexible substrate by aqueous method. From Figure 2a, a large of ZnO nanorod array grew uniformly on the flexible substrate with varying diameter of nanorods (80–100 nm). The shapes of the nanorods had wurtzite structure. The inset of the Figure 2a shows the cross-section of the nanorods. As seen, most of the nanorods were well aligned, perpendicular to the substrate which the length of the nanorods can reach up to approximately 1 μm. Figure 2b shows the top view of the ZnO MSM nanorod device. From the image, it can be seen that the ZnO nanorod arrays selectively distributed among the interdigitated electrodes of the device. It confirmed that the electrodes can effectively be protected by the photoresist, which avoid the nanorods growing onto the electrodes. Therefore, our method can fabricate MSM PDs with a large area aligned ZnO nanorods to enhance the performance.

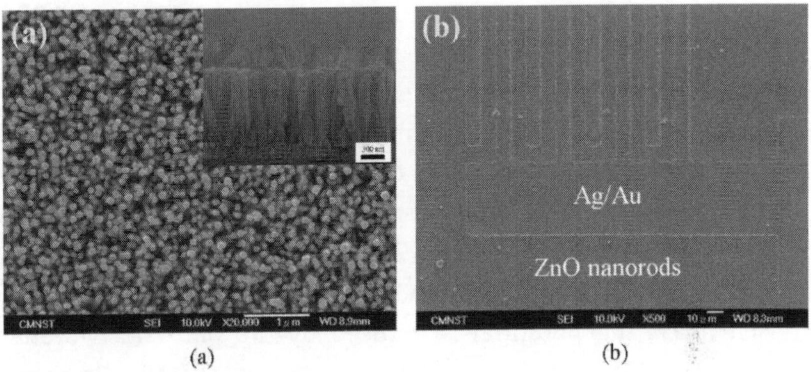

Figure 2 SEM images. (**a**) Top view of the ZnO nanorods. Inset: the cross-sectional image of the nanorods. (**b**) Top view of the ZnO nanorods MSM photodetector.

Figure 3 shows the typical X-ray diffraction spectrum of the ZnO nanorods prepared on the ZnO/PI substrate. According to XRD result, it depicts the diffraction angles of 34.4° assigned as ZnO (0002) reflection. The strong (0002) reflection with narrow width indicates that the ZnO nanorods were crystallized in wurtzite structure and preferentially grown along the *c*-axis direction.

The ZnO nanorod MSM PDs were fabricated on flexible polyimide substrate, so the performance under bending conditions is of major importance. It is important to understand the characteristics of the flexible ZnO nanorod MSM PDs with bending and without bending for the future of plastic electronic productions. Stable performance at bending situation is a critical challenge. For this reason, we measured the electrical characteristics of the PDs with bending and without bending. Before the measurement, we prepared a holder with semicircular shape (the radius of the holder was 0.2 cm) and fixed the ZnO nanorod MSM PDs on it, which could make the device bending. For convenience of realization, Figure 4a shows a schematic drawing of the bending device, where *r* corresponds to the radius of the curvature of bending. A smaller radius of the curvature of bending means increased bending. Figure 4b shows the current–voltage characteristics of ZnO nanorod PDs under flat and 0.2-cm radius of bending curvature, respectively. From the result, it could be known that dark current decreased slightly under bending, but the shift was lesser than one order of magnitude. The I-V behavior of the device was modulated due to the change in Schottky barrier height (SBH) at the metal–semiconductor interface. It is well known that the lacking of center symmetry in ZnO lead ionic polarization can be induced by strain [29]. When the ZnO nanorod PDs are bent by an external force, the potential is induced by the piezoelectric effect which is created by the relative displacement of the Zn^{2+} with respect to the O^{2-}, and these piezoelectric ionic charges cannot move freely until external force releases [21, 30]. These piezoelectric ionic charges could affect the charge transport strongly. The changing of piezopotential would shift the local Fermi level and modify the local conduction band structure. Therefore, the PDs under tensile strain would increase the SBH and influence the charge transport property by the piezoelectric effect and the band structure.

In our experiment, the structure of our device is MSM structure which is like two back-to-back Schottky diodes, and our devices were measured at room temperature and the ZnO nanorods had low doping. Therefore, the current transport mechanism followed the thermionic emission-diffusion theory (for $V >> 3\ kT/q$, approximately 77 mV) [31]. According to the literature [23], the thermionic emission-diffusion theory can be simplified by assuming some parameters are independent of strain, and then the change of SBH can be determined by the equation below [23]:

$$\ln[I(\epsilon_{zz})/I(0)] = -\Delta\phi_s/kT \quad (1)$$

where I (ϵ_{zz}) and $I(0)$ are the current measured from the ZnO nanorod MSM PDs at a fixed bias with and without being strained, respectively. After calculation, the changes of SBH were 27 meV and 29 meV for two biases of 1 and 2 V, respectively. From the two values, it can be known that the change of the SBH is not very sensitive to the bias applied across the device.

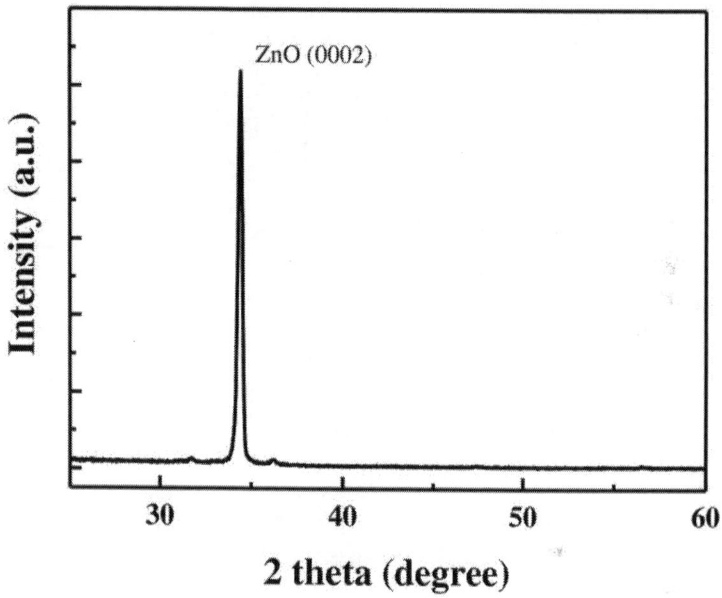

Figure 3 XRD spectrum measured from vertically aligned ZnO nanorods.

Owing to the outward bending of the substrate, piezoelectric effects could be induced and it could be attributed to the tensile strain between n-ZnO lattices. As we know, strain can be used in many applications. In microelectronic industry, scaling of MOSFETs has attracted many people to study it for the improvement in integrated circuit density and performance. However, MOSFET size reduction has met technological challenges such as short-channel effect and high leakage current. Therefore, strain has been an increase in interest in semiconductor material since the 1950s. Using appropriate tensile and compressive strain components in the n- and p-MOSFET's channel region could enhance its performance [32–35].

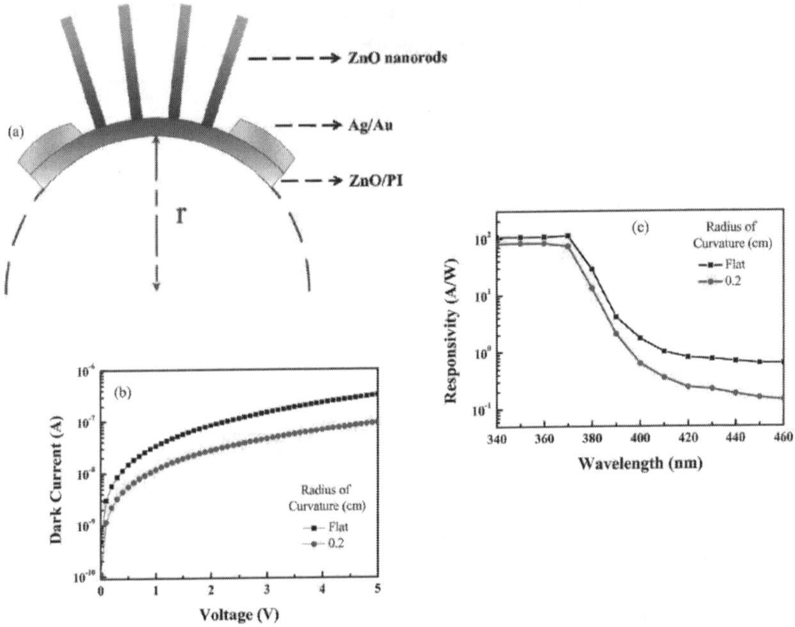

Figure 4 Schematic drawing of the bending device, and the dark I-V characteristics and responsivity characteristics. (**a**) Schematic drawing of the bending device. (**b**) The I-V and (**c**) responsivity characteristics of the PD measured from flat and bending substrate in the dark.

Figure 4c shows the responsivity characteristics of the ZnO nanorod PDs under flat and bending curvature. In this figure, the responsivities in visible wavelengths decreased slightly under bending, and the cutoff wavelength was the same. The UV-to-visible rejection ratios, defined as the ratio of the responsivity at 370 nm and at 460 nm, are 174.6 and 503.2 for flat and 0.2-cm radius of curvature bending, respectively. The rejection ratio of the 0.2-cm radius of curvature bending had a larger value that could be attributed to the smaller dark current.

From the responsivity characteristics of the ZnO nanorod PDs, the PDs had high responsivities under UV-light illumination. In addition, the photoconductive gain is given by the following equation [36]:

$$G = Rh\nu/q\eta \quad (2)$$

where G is the internal gain, R is the responsivity, h is the Plank's constant, ν is the frequency of the light, q is the electronic charge, and η is the quantum efficiency. Assuming η is 1 for simplicity, the internal gains of the ZnO nanorod PDs measured at flat and 0.2-cm radius of curvature bending were 3.71×10^2 and 2.44×10^2, respectively. Both PDs had high responsivities due to ZnO nanorod large surface-to-volume ratio and the existence of the oxygen-related hole surface states on the ZnO nanorods surface. The oxygen molecules can adsorb on the ZnO nanorod surface by capturing free electrons and desorb from the surface by illustrating UV light which lead to an increase in the free carrier concentration, a decrease in the width of the depletion layer, and a reduction of the Schottky barrier height [37]. Recently, Soci et al. [37] published the ZnO nanowire PDs with high internal gain. They fabricated a ZnO photodetector with one ZnO nanowire. Their device has achieved an internal gain of approximately 10^8. Instead, our device has smaller internal gain than theirs. This could be attributed to the high density of nanorod array of our device and there were small gaps between the freestanding nanorods. When the nanorods were illustrated by the UV light, the oxygen molecules desorbed from the nanorods surface. However, the nanorods were too close that make the oxygen molecules quickly readsorb with the nanorod surface by capturing free electrons and decrease the photocurrent. Thus, the internal gain of our device was smaller than theirs.

4. CONCLUSIONS

In summary, we reported the fabrication and I-V characteristics of ZnO nanorod PDs on PI substrate. The crystal and optical properties of ZnO nanorods had been investigated by SEM and XRD. They showed the ZnO nanorods grown on the PI substrate had high density array, hexagonal wurtzite structure and well (0002) crystalline phase. From I-V curve, it was found that the dark current decreased slightly and the UV-to-visible rejection ratio increased slightly in bending situation. The decreasing tendency of dark current under bending may be attributed to the increase of the Schottky barrier height. Consequently, ZnO nanorods have great potential for high-performance UV photodetectors, and polyimide is a promising substrate for flexible PDs and devices.

ACKNOWLEDGMENTS

This work was supported by National Science Council of Taiwan under contract number NSC 100-2221-E-150-057. National Formosa University Research and Services Headquarters that provided the partial equipment for measurement is also acknowledged.

COMPETING INTERESTS

The authors declare that they have no competing interests.

AUTHORS' CONTRIBUTIONS

TPC, SJY, and SJC designed the whole experimental procedure and related analysis. CHH participated in data analysis. YJH did the nanowire growth and other experiments. All authors read and approved the final manuscript.

REFERENCES

1. Hvam JM: Direct recording of optical-gain spectra from Zno. *J Appl Phys* 1978, 49: 3124–3126.

2. Look DC: Recent advances in ZnO materials and devices. *Mater Sci Eng: B* 2001, 80: 383–387.

3. Solis-Pomar F, Martinez E, Melendrez MF, Perez-Tijerina E: Growth of vertically aligned ZnO nanorods using textured ZnO films.*Nanoscale Res Lett* 2011, 6: 524.

4. Morkoc H, Strite S, Gao GB, Lin ME, Sverdlov B, Burns M: Large-band-gap sic, III-V nitride, and II-VI ZNSE-based semiconductor-device technologies.*J Appl Phys* 1994, 76: 1363–1398.

5. Anwar AFM, Wu SL, Webster RT: Temperature dependent transport properties in GaN, A1(x)Ga(1-x)N, and InxGa1-xN semiconductors.*IEEE T Electron Dev* 2001, 48: 567–572.

6. Look DC, Reynolds DC, Sizelove JR, Jones RL, Litton CW, Cantwell G, Harsch WC: Electrical properties of bulk ZnO. *Solid State Commun*1998, 105: 399–401.

7. Dietrich CP, Lange M, Stolzel M, Grundmann M: Microwire (Mg,Zn)O/ZnO and (Mg,Zn)O/(Cd,Zn)O non-polar quantum well heterostructures for cavity applications. *Appl Phys Lett* 2012, 100: 031110–1-031110–4.

8. Zhang J, Zhao HP, Tansu N: Large optical gain AlGaN-delta-GaN quantum wells laser active regions in mid-and deep-ultraviolet spectral regimes.*Appl Phys Lett* 2011, 98: 171111–1-171111–3.

9. Zhao HP, Liu GY, Zhang J, Poplawsky JD, Dierolf V, Tansu N: Approaches for high internal quantum efficiency green InGaN light-emitting diodes with large overlap quantum wells. *Opt Express* 2011, 19: A991-A1007.

10. Shukla G: ZnO/MgZnO p-n junction light-emitting diodes fabricated on sapphire substrates by pulsed laser deposition technique. *J Phys D: Appl Phys* 2009, 42: 075105.

11. Ahn J, Park H, Mastro MA, Hite JK, Eddy CR, Kim J: Nanostructured n-ZnO/thin film p-silicon heterojunction light-emitting diodes. *Opt Express* 2011, 19: 26006–26010.

12. Leung YH, He ZB, Luo LB, Tsang CHA, Wong NB, Zhang WJ, Lee ST: ZnO nanowires array p-n homojunction and its application as a visible-blind ultraviolet photodetector. *Appl Phys Lett* 2010, 96: 053102.

13. Sun F, Shan C-X, Wang S-P, Li B-H, Zhang Z-Z, Yang C-L, Shen D-Z: Ultraviolet photodetectors fabricated from ZnO p-i-n homojunction structures. *Mater Chem Phys* 2011, 129: 27–29.

14. Kim J, Yun J-H, Kim CH, Park YC, Woo JY, Park J, Lee J-H, Yi J, Han C-S: ZnO nanowire-embedded Schottky diode for effective UV detection by the barrier reduction effect. *Nanotechnology* 2010, 21: 115205.

15. Qin L, Shing C, Sawyer S: Metal–semiconductor–metal ultraviolet photodetectors based on zinc-oxide colloidal nanoparticles. *IEEE Electron Device Lett* 2011, 32: 51–53.

16. Yen T, Yun J, Kim SJ, Cartwright A, Anderson WA: Photocurrent enhancement in nanocrystalline-ZnO/Si heterojunction metal–semiconductor–metal photodetectors. *Electrochemical Solid-State Lett* 2011, 14: H415-H418.

17. Li D, Sun X, Song H, Li Z, Chen Y, Miao G, Jiang H: Influence of threading dislocations on GaN-based metal–semiconductor–metal ultraviolet photodetectors. *Appl Phys Lett* 2011, 98: 011108.

18. Yu J, Ippolito SJ, Wlodarski W, Strano M, Kalantar-Zadeh K: Nanorod based Schottky contact gas sensors in reversed bias condition.*Nanotechnology* 2010, 21: 265502.

19. Han JB, Fan FR, Xu C, Lin SS, Wei M, Duan X, Wang ZL: ZnO nanotube-based dye-sensitized solar cell and its application in self-powered devices. *Nanotechnology* 2010, 21: 405203.

20. Wang ZL: ZnO nanowire and nanobelt platform for nanotechnology. *Mat Sci Eng R* 2009, 64: 33–71.

21. Wang ZL, Song JH: Piezoelectric nanogenerators based on zinc oxide nanowire arrays. *Science* 2006, 312: 242–246.

22. Kwon SS, Hong WK, Jo G, Maeng J, Kim TW, Song S, Lee T: Piezoelectric effect on the electronic transport characteristics of ZnO nanowire field-effect transistors on bent flexible substrates. *Adv Mater* 2008, 20: 4557–4562.

23. Zhou J, Gu YD, Fei P, Mai WJ, Gao YF, Yang RS, Bao G, Wang ZL: Flexible piezotronic strain sensor. *Nano Lett* 2008, 8: 3035–3040.

24. Wang ZL, Lao CS, Kuang Q, Park MC, Deng YL: Polymer functionalized piezoelectric-FET as humidity/chemical nanosensors. *Appl Phys Lett* 2007, 90: 262107.

25. Lee C-H, Kim Y-J, Hong YJ, Jeon S-R, Bae S, Hong BH, Yi G-C: Flexible inorganic nanostructure light-emitting diodes fabricated on graphene films. *Adv Mater* 2011, 23: 4614–4619.

26. Chu S, Li DD, Chang PC, Lu JG: Flexible dye-sensitized solar cell based on vertical ZnO nanowire arrays. *Nanoscale Res Lett* 2011, 6: 38.

27. Hwang JO, Lee DH, Kim JY, Han TH, Kim BH, Park M, No K, Kim SO: Vertical ZnO nanowires/graphene hybrids for transparent and flexible field emission. *J Mater Chem* 2011, 21: 3432–3437.

28. Ahsanulhaq Q, Umar A, Hahn YB: Growth of aligned ZnO nanorods and nanopencils on ZnO/Si in aqueous solution: growth mechanism and structural and optical properties. *Nanotechnology* 2007, 18: 115603.

29. Wang ZL, Wu WZ: Piezotronic nanowire-based resistive switches as programmable electromechanical memories. *Nano Lett* 2011, 11: 2779–2785.

30. He JH, Hsin CL, Liu J, Chen LJ, Wang ZL: Piezoelectric gated diode of a single ZnO nanowire. *Adv Mater* 2007, 19: 781–784.

31. Sze SM: *Physics of Semiconductor Devices*. 2nd edition. Wiley, New York; 1981.

32. Chu M, Sun Y, Aghoram U, Thompson SE: Strain: a solution for higher carrier mobility in nanoscale MOSFETs. *AnnuRev Mater Res* 2009, 39: 203–229. 10.1146/annurev-matsci-082908-145312

33. Lee MH, Chang ST, Wu TH, Tseng WN: Driving current enhancement of strained Ge (110) p-type tunnel FETs and anisotropic effect.*Electron Device Lett, IEEE* 2011, 32: 1355–1357.

34. Yeo YC: Enhancing CMOS transistor performance using lattice-mismatched materials in source/drain regions. *Semicond Sci Tech* 2007, 22: S177-S182.

35. Ghani T, Armstrong M, Auth C, Bost M, Charvat P, Glass G, Hoffmann T, Johnson K, Kenyon C, Klaus J, McIntyre B, Mistry K, Murthy A, Sandford J, Silberstein M, Sivakumar S, Smith P, Zawadzki K, Thompson S, Bohr M: A 90 nm high volume manufacturing logic technology featuring novel 45 nm gate length strained silicon CMOS transistors. In Electron Devices Meeting Tech Dig 2003, 11.6.1–11.6.3.

36. Bube RH: *PhotoconductiVity of Solids*. Wiley, New York; 1960:461.

37. Soci C, Zhang A, Xiang B, Dayeh SA, Aplin DPR, Park J, Bao XY, Lo YH, Wang D: ZnO nanowire UV photodetectors with high internal gain.*Nano Lett* 2007, 7: 1003–1009.

3

PROGRESS TOWARDS AN AUTONOMOUS FIELD DEPLOYABLE DIODE-LASER-BASED DIFFERENTIAL ABSORPTION LIDAR (DIAL) FOR PROFILING WATER VAPOR IN THE LOWER TROPOSPHERE

Kevin S. Repasky [1,*], Drew Moen [1], Scott Spuler [2], Amin R. Nehrir [3] and John L. Carlsten [4]

[1] Electrical and Computer Engineering, Montana State University, Bozeman, MT 59717, USA

[2] National Center for Atmospheric Research, Boulder, CO 80307, USA

[3] NASA Langley Research Center, Hampton, VA 23681, USA

[4] Physics Department, Montana State University, Bozeman, MT 59717, USA

ABSTRACT

A laser transmitter has been developed and incorporated into a micro-pulse differential absorption lidar (DIAL) for water vapor profiling in the lower troposphere as an important step towards long-term autonomous field operation. The laser transmitter utilizes two distributed Bragg reflector (DBR) diode lasers to injection seed a pulsed tapered semiconductor optical amplifier (TSOA), and is capable of producing up to 10 μJ of pulse energy with a 1 μs pulse duration and a 10 kHz pulse repetition frequency. The on-line wavelength of the laser transmitter can operate anywhere along the water vapor absorption feature centered at 828.187 nm (in vacuum) depending on the prevailing atmospheric conditions, while the off-line wavelength operates at 828.287 nm. This laser transmitter has been incorporated into a DIAL instrument utilizing a 35.6 cm Schmidt-Cassegrain telescope and fiber coupled avalanche photodiode (APD) operating in the photon counting mode. The performance of the DIAL instrument was demonstrated over a ten-day observation period. During this observation period, data from radiosondes were used to retrieve water vapor number density profiles for comparisons with the number density profiles retrieved from the DIAL data.

KEYWORDS

DIAL; trace gas sensing; ground-based lidar

1. INTRODUCTION

The planetary boundary layer is the bottommost part of the troposphere and typically ranges from the Earth's surface to several hundred meters, during stable nighttime conditions, to several kilometers, during the day, when solar induced convection causes shear and turbulence in the air parcel near the surface. Most of the convective processes that drive the Earth's weather patterns initiate within the planetary boundary layer where most of the atmospheric water vapor and aerosols reside [1,2]. Understanding the state of the atmosphere requires knowledge of the spatial distribution of water vapor in the lower troposphere [3]. Research in the development of water vapor profiling tools to meet this needed measurement capability of water vapor distribution in the lower troposphere is ongoing [4,5].

Active optical remote sensing can be used for range resolved water vapor profile measurements in the lower troposphere. Two laser based active remote sensing techniques for water vapor profiling include Raman lidar [6–11] and differential absorption lidar (DIAL) [12–20]. Raman lidar uses the characteristic frequency shift resulting from the inelastic Raman scattering to identify the molecular species of interest while DIAL utilizes the elastically scattered signals from two closely spaced laser transmitter wavelengths, the first associated with an absorption line for the molecule of interest, referred to as the on-line wavelength, and the second minimally affected by any molecular absorption, referred to as the off-line wavelength. The difference in the strength of the on-line and off-line return signals results from molecular absorption so that with knowledge of the absorption cross section, the ratio of scattered light collected by the DIAL receiver at the on-line and off-line wavelengths can be used to determine the molecular number density.

Modeling of low pulse energy, high pulse repetition rate DIAL instruments indicated that with the available avalanche photodiodes (APDs) operating in the photon counting Geiger mode, micro-pulse DIAL instruments are feasible with adequate spatial resolution and averaging times needed for monitoring water vapor in the lower troposphere [21,22]. The advantages of the diode-based micro-pulse DIAL include the ability to develop low-cost eye-safe instruments for autonomous network deployment, modularity, small footprint, and high wall plug efficiency. Recognizing the potential of the micro-pulse DIAL technique, Machol *et al.* [15] developed a diode-laser-based DIAL instrument based on distributed feedback (DFB) diode lasers and tapered semiconductor optical amplifiers (TSOAs) and demonstrated initial water vapor retrievals up to 2.5 km during nighttime operation. Building on the initial work of Machol *et al.*, researchers at Montana State University have developed a micro-pulse DIAL based on two external cavity diode lasers (ECDL's) and a single pulsed TSOA [16–20]. This instrument has retrieved daytime water vapor profiles up to 4 km and nighttime water vapor profiles up to 6 km, providing the first demonstration of a diode-laser-based micro-pulse DIAL instrument capable of full-time water vapor and aerosol retrievals. However, the ECDL's used in this instrument were prone to mode hopping inhibiting long term autonomous operation.

One major technological improvement over previously demonstrated instruments that is needed is the development of a stable laser transmitter, both spectrally and mechanically, capable of operating over long periods of time with minimal user interactions as an important step towards long term autonomous field deployment of micro-pulse DIAL instruments. A diode-laser-based micro-pulse DIAL is described in this manuscript that addresses improvements to the laser transmitter. The major improvement to the laser transmitter includes the elimination of the ECDL's used by Nehrir *et al.* [16–20], which would experience mode hops resulting from temperature changes in the local operating environment causing a loss in the frequency stabilization needed for long term DIAL operation. The ECDLs have been replaced by distributed Bragg reflector (DBR) diode lasers that are more stable against mechanical and thermal disturbances. Furthermore, many free space optical elements used in the laser transmitter have been replaced by fiber-coupled components as a further step in reducing the thermal and mechanical disturbances on optical alignment. To demonstrate the improvements resulting in the aforementioned hardware changes, the DIAL system was operated continuously during a ten-day observation period between 17 and 27 July 2013. The data products retrieved from the DIAL instrument during this observation period include the normalized relative backscatter and water vapor number density [19]. During this observational period, Vaisala RS92 radiosondes were launched providing temperature, pressure, and relative humidity profiles. Comparisons between the retrieved DIAL and radiosonde number density profiles were used to validate the DIAL measurements.

This paper is organized as follows: a description of the DIAL instrument is given in Section 2, a discussion of the data collected with the DIAL instrument is presented in Section 3, and concluding remarks are made in section 4.

2. INSTRUMENT

2.1. Laser Transmitter

A schematic of the DIAL instrument is shown in Figure 1. The laser transmitter utilizes two DBR lasers mounted in a TO-8 package (manufactured by Photodigm, part number PH828DBR100T8) operating with a nominal wavelength of 828.2 nm. The laser linewidth was measured using a heterodyne technique based on mixing of two DBR laser to produce a beat note with a full width half maximum value of measured linewidth less than 1 MHz (0.023 pm). Each DBR laser can output up to 80 mW of continuous wave (cw) optical power and can be thermally tuned with a tuning response of 0.059 nm/C or tuned by adjusting the drive current with a tuning response of 0.0011 nm/mA. Each DBR laser is controlled using a commercial current and temperature controller (ILX Lightwave LDC3724C). The output from each DBR laser is collimated using an aspheric lens with a focal length of 4.0 mm and numerical aperture of 0.6. After the collimating lens, the cw light passes through a Faraday isolator (Thor Labs IO-3D-830-VLP) to prevent unwanted feedback and is then incident on a half-wave-plate to align the state of polarization to the polarization maintaining fibers. The cw light is then coupled into a single mode polarization maintaining fiber with angled physical contact (APC) connectors using an 11.0 mm focal length lens with a numerical aperture of 0.22. The polarization maintaining fiber delivers the light to a polarizartion maintaining 90/10 in-line fiber splitter (Newport F-PMC-850-10) with the 10% output port used for wavelength monitoring purposes while the 90% output port from each of the two laser assemblies is directed to a 2 ×

1 fiber coupled micro-electro-mechanical systems (MEMS) switch (Aglitron LBPM-128238333).

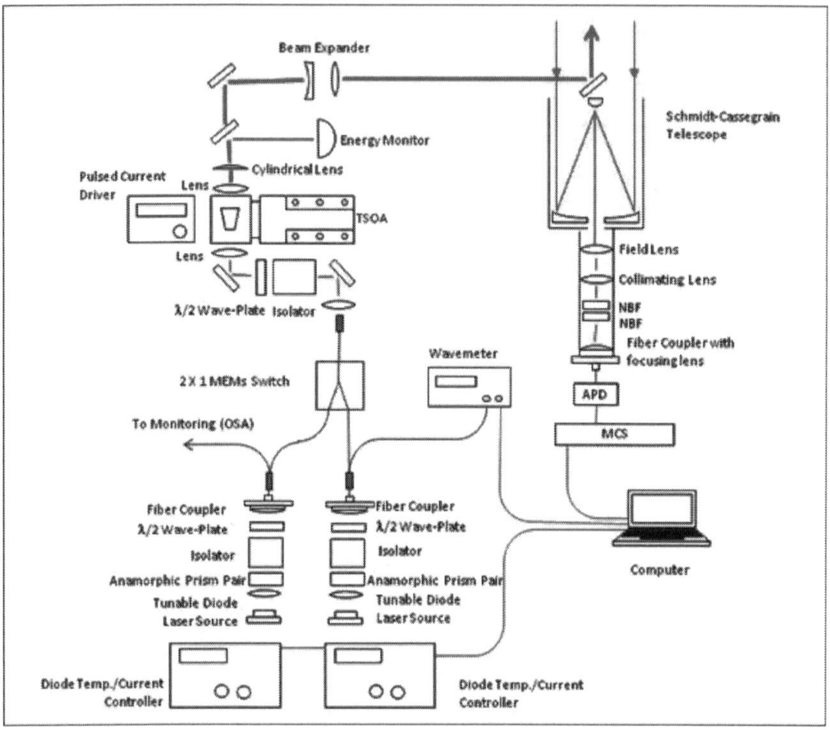

Figure 1. Schematic of the diode laser based DIAL instrument. The Laser transmitter is based on two DBR lasers used to injection seed a TSOA in a MOPA configuration.

The switch allows the injection seeding of the tapered semiconductor optical amplifier (TSOA) (Eagleyard EYP-TPA-830-01000-4006-CM-TO4) with either the on-line or off-line wavelength. Fiber coupled light from the switch is collimated using a lens with a focal length of 4.6 mm and a numerical aperture of 0.53 and next passes through a Faraday isolator to prevent unwanted feedback from the backward traveling amplified spontaneous emission from the TSOA entrance aperture to the seed lasers. After the Faraday isolator, the light is coupled into a 4.0 mm long ridge waveguide TSOA using a 4.5 mm focal length aspheric lens. The optical power of the seed laser for the on-line (off-line) wavelength right before the TSOA was 12.6 mW (13.9 mW). This seed power saturated the gain of the TSOA so that small changes in the input seed power did not appreciable affect the pulsed output energy of the TSOA. The output from the TSOA is collimated using a spherical lens, with a focal length of 4.5 mm, placed after the TSOA and a cylindrical lens with a focal length of 50 mm placed 79 mm after the spherical lens. These two lenses are needed to collimate the output of the TSOA, which exits from an output aperture of 3 μm × 190 μm. After the collimating lenses, approximately 2% of the light is directed to a low bandwidth photo detector used to monitor the output pulse energy while the remainder of the light is directed through a 5x beam expanded where the light is re-collimated. The collimated output beam has a measured M^2 value of 2 in the vertical plane and 6 in the horizontal plane. The M^2 value of 2 × 6 results in a full angle beam divergence of 14.6 μrad × 43.8 μrad for the 3.6 cm diameter collimated output beam. The field of view (FOV) of the receiver, which is discussed in detail in Section 2.2 below, is calculated to be 224 μrad.

The water vapor absorption line used has a center wavelength of 828.187 nm (in vacuum), a line-strength of 1.64×10^{-23}cm^{-1}/(molecule·cm^{-2}), a full width at half maximum pressure broadened linewidth at sea level of γ_0 = 0.1896 cm^{-1} (2.1 pm, 0.9 GHz), a linewidth temperature dependence of α = 0.74, and a ground state transitional energy of E" = 212.1564 cm^{-1}[16–20,23]. The on-line DBR laser is stabilized with the aid of an optical wavemeter (EXFO-WA1500) using a feedback control loop in the data acquisition program. The wavemeter has a resolution of ±55 MHz (±0.125 pm) and a readout rate of 1 Hz. This resolution is larger than the laser linewidth of 1 MHz (0.23 pm) resulting in an uncertainty in the operating wavelength set by the resolution of the wavemeter. The on-line wavelength stabilization is achieved by reading the wavelength and comparing the DBR operating wavelength to a user defined set wavelength. If the operating wavelength varies from the set wavelength, the current to the DBR laser is adjusted to make small corrections to the DBR operating wavelength. The advantage for this method of frequency stabilization is that it allows operation of the on-line wavelength anywhere along the water vapor absorption feature, providing the user the ability to choose the absorption cross-section to match the prevalent atmospheric conditions. Ideally, the absorption cross section is set so that optical depth, τ, at the desired maximum altitude, r_{max}, is given by $\tau = \int_0^{r_{max}} \sigma(r', \lambda_{on})N(r')dr' \approx 1 - 2$ where σ(r, λ) is the absorption cross section and N(r) is the number density and the optical depth over a single range bin, Δτ, is given by $\Delta\tau = \sigma(r, \lambda_{on})N(r)\Delta r \approx 0.02 - 0.1$, where Δr is the range bin size [24–26]. The off-line wavelength is set by adjusting the operating temperature and current at the beginning of the data collection period. This off-line laser is monitored using an optical spectrum analyzer with 0.06 nm (26 GHz) resolution. No active feedback control is provided to the off-line DBR laser.

Pulsed operation of the DIAL laser transmitter is achieved by pulsing the current to the TSOA. A commercial current driver capable of producing up to 15.7 A pulses up to 5 ms in pulse duration and up to 1 MHz pulse repetition frequency are used (Directed Energy, Inc, PCX-7420). The rise and fall time associated with the current pulse is 25 ns. For the DIAL operation, the pulse current driver was operated at 12.5 A with a 1 μs pulse duration and a 10 kHz pulse repetition frequency yielding a nominal pulse energy of 10.3 μJ. A summary of the laser transmitter parameters is given in Table 1.

Table 1. A summary of the laser transmitter parameters. The short-term linewidth resulted from the heterodyne measurement, while the long-term bandwidth results from the resolution of the wavemeter used to lock the on-line seed laser.

Parameter	Measured
Laser Seeder	2 DBR diode lasers
Amplifier	Single Stage TSOA
Transmitter Wavelengths	828.187 nm (on-line) 828.1965–828.2000 nm (side-line) 828.287 nm (off-line)
Pulse Duration	1 μs
Pulse Repetition Rate	10 kHz
Pulse Energy	10 μJ
Parameter	**Measured**
Short Term Linewidth	<1 MHz (0.023 pm)
Long Term Bandwidth	±55 MHz ± (0.125 pm)
Beam Diameter	3.8 cm
Switching Time	6 s
Averaging Time	20 min

2.2. DIAL Receiver

The receiver design described here closely follows the design previously described by Nehrir *et al.* [18,19]. Light scattered in the atmosphere is collected using a 35.56 cm diameter F/11 Schmidt-Cassegrain telescope. At the back end of the telescope, an adaptor was made to couple a lens tube assembly housing the receiver optics to the telescope. A field lens with a 2.54 cm diameter and a focal length of 7.5 cm was placed at the focal plane of the telescope. A collimating lens with a 2.54 cm diameter and focal length of 5 cm was placed 5 cm behind the field lens producing a nominally collimated beam with a beam diameter of 4.5 mm. The collimated light next passes through two narrowband interference filters. Each filter has a maximum transmission of $T_0 = 0.65$, a center wavelength of $\lambda_0 = 828.206$ nm, a half-width-half-maximum of $\Delta\lambda = 125$ pm. After the narrowband filters, a fiber coupler with a 4.6 mm clear aperture and a 1.1 cm focal length is used to couple the light into a step index optical fiber with a 105 μm diameter and a numerical aperture of 0.22, which acts as the receiver field stop. This is discussed in more detail below where a model of the optical receiver is developed.

Lidar instruments rely on the efficient delivery of the collected backscattered light to the detector. Micro-pulse lidar and micro-pulse DIAL instruments face the added challenge of minimizing the background light seen by the detector so that the small signal resulting from the low pulse energy can overcome the background signal. The DIAL receiver described in this paper utilizes an optical fiber to maintain a narrow field of view (FOV) to minimize the background light delivered to the detector. The design trade-off with maintaining the narrow FOV to try to minimize the background signal is the range at which the DIAL comes in to full overlap. The DIAL receiver design tries to strike a balance of maintaining the narrow FOV while achieving full overlap at an acceptable range.

A model of the DIAL receiver was developed to estimate the receiver performance in terms of overlap and FOV. The model is based on a ray-tracing analysis using the ABCD matrices [27–29]. This analysis keeps track of the perpendicular distance, R, and slope, R′, of a ray relative to the optical axis using the vector:

$$R_a = \begin{pmatrix} R \\ R' \end{pmatrix} \tag{1}$$

Each optical element is characterized by a 2 × 2 matrix. For the two optical elements needed for modeling the receiver, the 2 × 2 matrices are given by Equation (2a) for a lens, and Equation (2b) for free space propagation as:

$$M_l = \begin{pmatrix} 1 & 0 \\ -1/f & 1 \end{pmatrix} \tag{2a}$$

$$M_{fs} = \begin{pmatrix} 1 & d \\ 0 & 1 \end{pmatrix} \tag{2b}$$

where f is the focal length of the lens and d is the free space propagation distance. The optical system is characterized by the 2 × 2 matrix $M = M_n\ M_{n-1} \cdots M_2 M_1$ where the subscript represents the order in which the optical element is encountered by the propagating ray. The output ray is then characterized by the vector R_b where $R_b = MR_a$ with the vector R_a characterizing the input ray.

The ray tracing analysis, outlined above, was used to model the optical receiver. This modeling assumes an outgoing beam radius of 2 cm with a full angle beam divergence of 25 μrad. The fiber, which acts as the field stop for the receiver, is located at the sum of the fiber coupling lens focal length plus an offset behind the fiber coupling lens. A plot of the beam radius at the fiber location is shown in Figure 2 as a function of range for various offsets. Overlap is achieved for the DIAL receiver when the beam radius is less than the fiber optic radius. For the optical receiver setup, light scattered from infinity will be imaged 1.1 cm behind the fiber coupling lens corresponding to a zero offset. However, as seen in Figure 2, overlap will not be achieved until a range of 4.1 km. By adjusting the offset, the range at which overlap is achieved can be reduced, however, if the offset is increased too far, the beam radius will become bigger than the fiber radius at larger ranges. The DIAL receiver offset was set at approximately 200 μm allowing full overlap to be achieved at 2.1 km while allowing the beam diameter to stay smaller than the fiber diameter up to 12 km.

For efficient coupling into the optical fiber, the incoming ray angles need to be less than the angle defined by the numerical aperture of the optical fiber. A plot of the maximum angle after the fiber lens as a function of range is shown inFigure 3. For the optical fiber used in the DIAL receiver, the maximum ray angle needs to be less than 0.22 radians, which, as seen from Figure 3, is achieved up to 12 km.

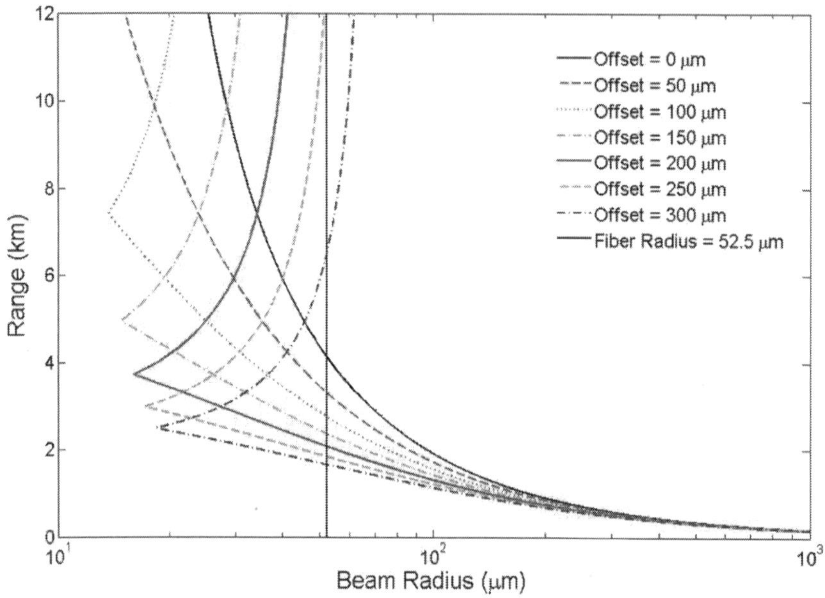

Figure 2. The beam radius at the optical fiber as a function of range and offset. The fiber radius is 52.5 μm as is shown as the vertical solid black line. Overlap occurs when the beam radius is less than the fiber radius.

The field of view of the DIAL receiver is set by the optical fiber, which acts as the receiver field stop. The field of view is calculated by propagating rays from the optical fiber through the receiver optical elements using the ray propagation model described above. The full angle field of view is calculated to be 224 μrad. This narrow field of view is important as it sets the amount of background light that will be collected and delivered to the detector providing the main source of noise.

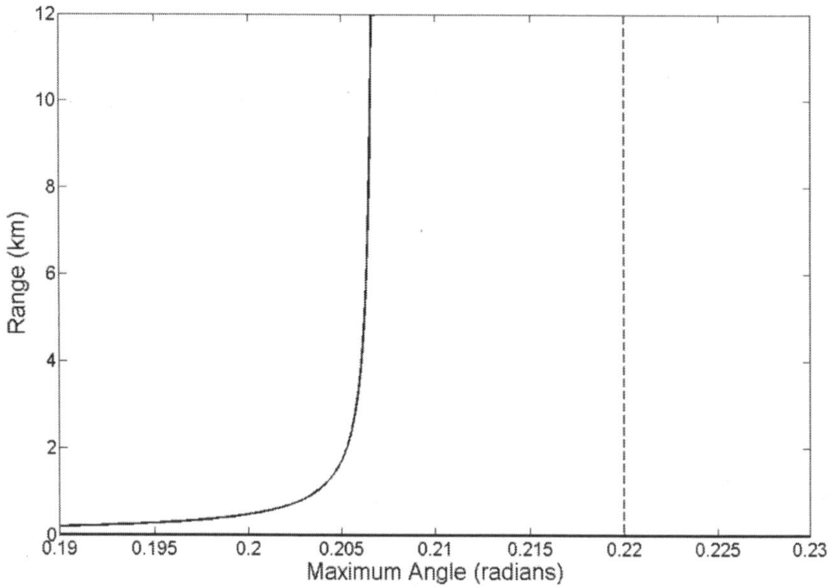

Figure 3. The maximum ray angle at the optical fiber as a function of range. The fiber numerical aperture requires angles less than 0.22 radians for efficient coupling. The vertical red dashed line indicated the maximum ray angle accepted by the optical fiber.

The receiver is monitored using a fiber coupled APD (Ecelitas SPCM-AQRH-13-FC) and a 20 MHz 4 channel multichannel scaler card (MCS) (Sigma Space AMCS-USB-4 Channel) data acquisition system. The laser transmitter generates pulses of 1 μs in duration corresponding to a range bin of 150 m; however, the MCS operates at 20 MHz, it collects a sample every 50 ns corresponding to a sampled range bin resolution of 7.5 m. Thus, the MCS collects 20 data points per 150 m pulse. During the post processing, 20 sampled range bins are summed to yield the 150 m range resolution of the DIAL instrument. A summary of the DIAL receiver specifications is given in Table 2.

Table 2. A summary of the differential absorption lidar (DIAL) receiver parameters.

Parameter	Measured
Telescope	Schmidt-Cassegrain
Primary Mirror Diameter	35.56 cm
Full Angle Field of View	224 μrad
Detector	Si Photon Counting APD
APD Quantum Efficiency	45%
Optical Filter Bandwidth	250 pm
Range Resolution	150 m

2.3. Data Collection

The instrument is operated and data collected using the Labview programming environment. The on-line wavelength is actively locked by querying the optical wavemeter every 1 s and comparison of the operating wavelength of the DBR laser with a user defined operating wavelength allows adjustments to the on-line wavelength by varying the operating current of the on-line DBR laser. The MEMS switch is then set so that the on-line DBR laser seeds the TSOA and data collected over a user defined number of sampled range bins is averaged over 6 s using the MCS. The control program then reads the MCS and stores the data as a row in the on-line data matrix. Next, the MEMS switch is changed so that the off-line DBR laser is seeding the TSOA and the process is repeated with the data being stored in the off-line data matrix. The 6 s averaging time at each wavelength was chosen as a compromise between trying to capture rapidly changing atmospheric conditions and data transfer between the MCS and computer.

Data are processed in the matlab programming environment. First, the count rate for each range bin is estimated and the counting correction factor, given by $T_c = 1/(1 - t_d N)$ where t_d = 39.2 ns is the characteristic dead time associated with the APD, is applied for both the on-line and off-line return signals. Then, the last 50 sampled range bins are averaged for each return profile and this average value is used for background subtraction. The off-line wavelength returns are then used to generate the normalized relative backscatter profile using a 1 min averaging time while the on-line and off-line data are used to calculate the water vapor number density profile through application of the DIAL equation using a 20 min averaging time (10 min at each wavelength). The normalized relative backscatter retrieval uses a 150 m spatial averaging while the water vapor number density retrieval uses a 150 m spatial averaging below 3 km and a 300 m spatial averaging above 3 km. A flow chart summarizing the data processing is shown in Figure 4.

3. OBSERVATIONAL DATA

The DIAL instrument was operated over a ten-day period starting at 00:00 Mountain Daylight Time (MDT) 17 July 2013 and ending at 08:00 MDT 27 July 2013 at Montana State University. The DIAL operated in a laboratory environment with a major modification to the roof-port hatch allowing the installation of window for continuous operation regardless of meteorological conditions. The window has a 41 cm diameter and is not antireflection coated for the operating wavelength of the DIAL instrument coated. Measurement of the return signal with and without the window indicated that the window decreased the return signal by approximately 20% and will affect the signal to noise performance of the DIAL.

At the beginning of this observational period, storms occurred in western Montana as a frontal system moved through the area. By 19 July 2013, this frontal system moved out of the area and a high pressure system dominated resulted in clear skies. By 25 July 2013, another frontal system moved into the area for the remainder of the observation period.

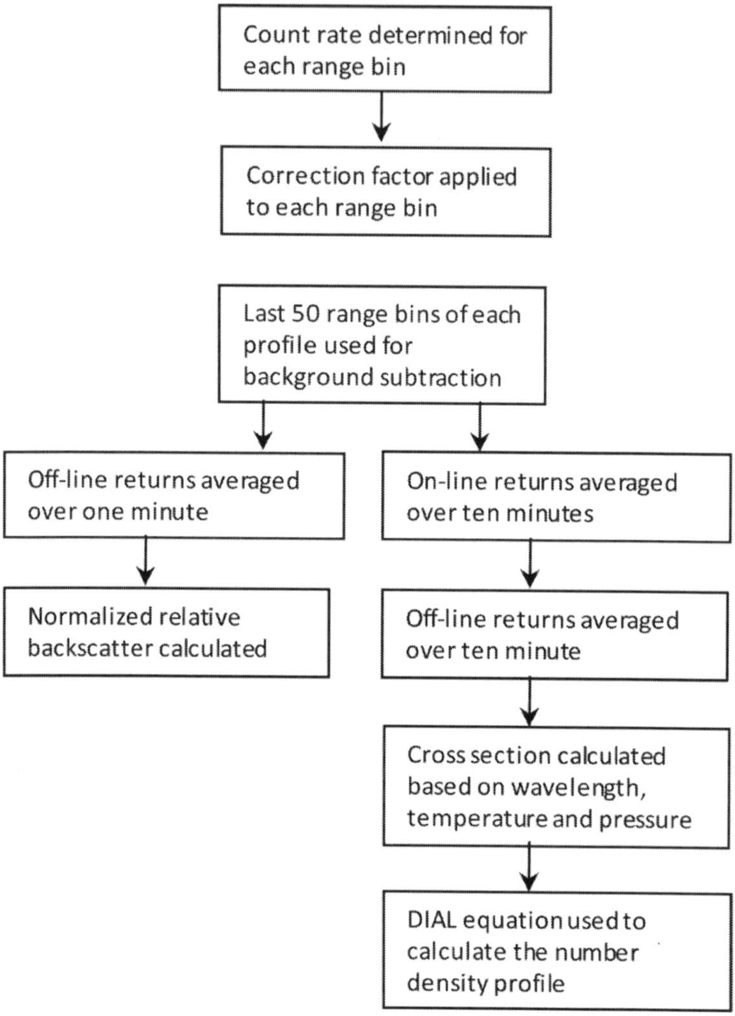

Figure 4. A flow chart summarizing of the data processing.

The main data products for the micro-pulse DIAL include the normalized relative backscatter and the water vapor number density. The normalized relative backscatter provides qualitative information on the aerosol distribution and a plot of the normalized relative backscatter as a function of range and time is shown as the upper plot in Figure 5 for the ten-day observational period. Because the DIAL does not achieve full overlap until 2.1 km, an overlap correction factor needs to be applied for accurate retrieval of the normalized relative backscatter below 2.1 km. The correction factor can be determined experimentally by operating the instrument along a horizontal path or through modeling by utilizing the matrix derivation of the optical transfer function as described above. The aerosol optical depth measured by a co-located solar radiometer as part of the NASA AERONET program is shown in the bottom plot in Figure 5 for reference. The aerosol optical depth measured by AERONET ranges between 0.02 and 0.10 over the ten-day observational. The higher optical depths during this observational period most likely resulted from smoke from forest fires burning in the Western United States.

The water vapor number density as a function of time and range is shown as the upper plot in Figure 6 with the white vertical lines indicating radiosondes launches. Using a co-located ground-based weather station, the temperature, relative humidity, and water vapor number density at the surface is shown in the lower three plots of Figure 6. The frontal system present 17–18 July 2013 and the weaker frontal system 25–27 July 2013 resulted in higher water vapor number densities as seen in the upper plot of Figure 5, as compared to number densities seen 19–24 July 2013 when a high-pressure system dominated resulting in clear skies. During this observational period, the laser transmitter was operated at 828.2000 nm 17–18 July 2013 when relatively high water vapor number densities were present. Operating at 828.2000 nm changed the pressure broadened absorption cross section from its peak value of 5.51×10^{-23} cm^2 at line center to 1.10×10^{-23} cm^2. With the high pressure system resulting in clear skies 19–24 July 2013, the operating wavelength was changed to 828.1965 nm–828.1975 nm resulting in a pressure broadened absorption cross section of 1.81×10^{-23} cm^2–1.53×10^{-23} cm^2. With the weaker frontal system moving through 25–27 July 2013, the operating wavelength was increased to 828.1975 nm–828.1985 nm resulting in a pressure broadened absorption cross section of 1.53×10^{-23} cm^2–1.34×10^{-23} cm^2. The change in the on-line operating wavelength in response to the changing meteorological conditions was incorporated to try to maintain the optical depth between 1 and 2 at 5 km as discussed in Section 2.1 above.

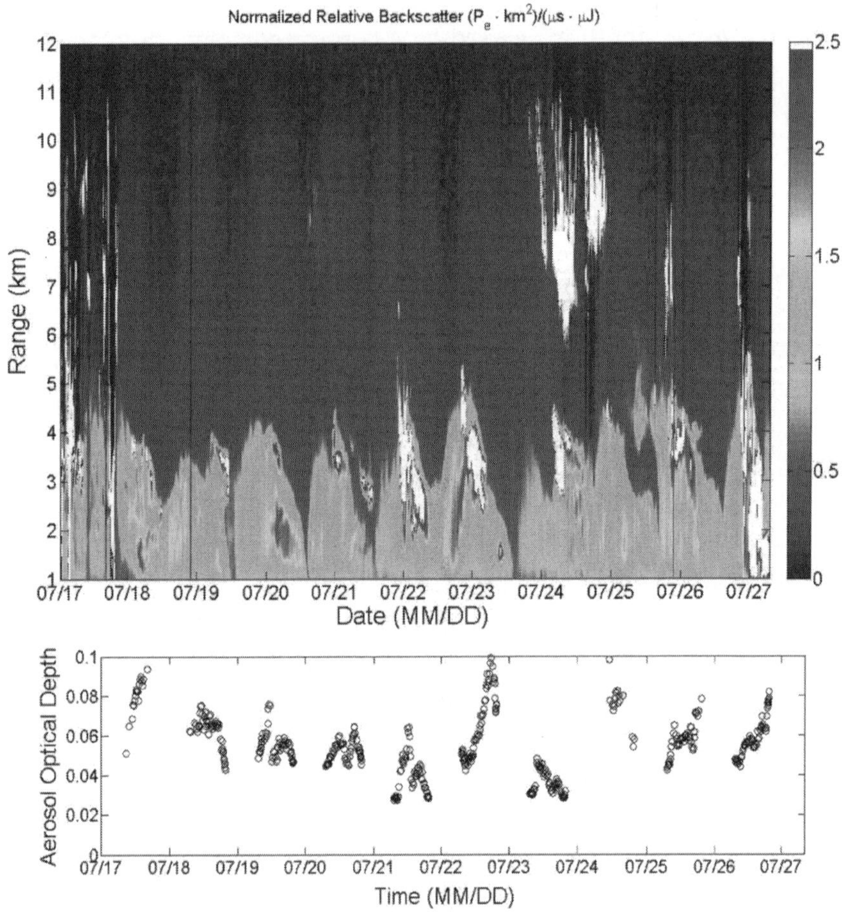

Figure 5. A plot of the normalized relative backscatter as a function of range and time is shown in the upper plot. The corresponding aerosol optical depth at 830 nm measured with a co-located solar radiometer as part of the NASA AERONET program is shown in the bottom plot.

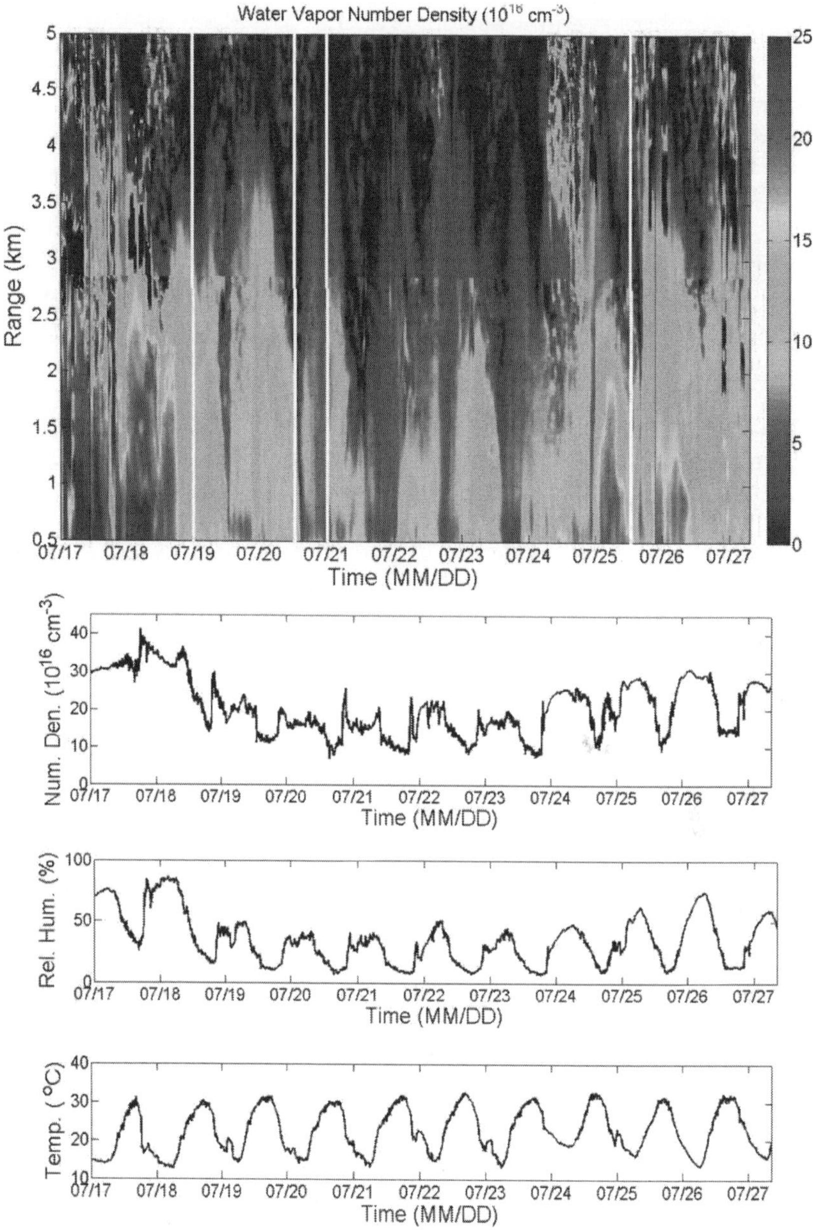

Figure 6. The water vapor number density as a function of range and time is shown in the top plot while the number density, relative humidity, and temperature measures at the surface are shown in the bottom three plots. The solid white vertical lines in the top plot indicate radiosonde launches.

During the observation period, Vaisala RS92 radiosondes were launched providing temperature, pressure and relative humidity profiles. A plot of the water vapor number density as a function of range for radiosondes launched 19 July 2013 at 00:16 MDT, 20 July 2013 at 21:55 MDT, 20 July 2013 at 12:45 MDT, and 25 July 2013 at 11:21 MDT are shown in the top-left, top-right, bottom-left, and bottom-right hand plots respectively in Figure 7. The red dashed line represents the number density retrieved from the temperature and relative humidity profiles measured by the radiosondes. The radiosonde profiles were filtered using a 20-point gliding filter corresponding to approximately a 60 m spatial average depending on the ascent rate of the radiosonde. The solid black line represents the DIAL retrieval at the time of the radiosonde launch. The error associated with the number density, Δn(r), is found as [30]:

$$\Delta n(r) = \frac{1}{2\left(\sigma_{on}(r) - \sigma_{off}(r)\right)\Delta r}\frac{1}{mk}\left\{\frac{1}{N(\lambda_{on}, r)} + \frac{1}{N(\lambda_{on}, r + \Delta r)} + \frac{1}{N\left(\lambda_{off}, r\right)} + \frac{1}{N\left(\lambda_{off}, r + \Delta r\right)}\right\}^{\frac{1}{2}} \quad (3)$$

where σ_{on} (r) (σ_{on} (r)) is the absorption on-line (off-line) absorption cross section at range r, m is the number of range bins averaged (m = 20 which corresponds to 150 m) and k is the number of samples in time (k = 100 which corresponds to one sample every six seconds averaged over 10 min), and N(λ,r) is the number of photons returned from range bin r at the wavelength λ. The grey shaded region represents the estimated error associated with the DIAL retrieval. The DIAL retrievals utilize a 20-minute averaging time and a 20-point gliding spatial filter corresponding to a 150 m spatial average. Good agreement between the radiosonde and DIAL profiles between 1.5 and 6 km for the two nighttime comparisons and between 1.5 and 4.5 km for the daytime comparisons is seen. Below 1.5 km, two factors can affect the DIAL retrieval including the range and wavelength dependent transmission of the narrow band optical filters [16,25] and the rapidly changing overlap function. For this micro-pulse DIAL, the pulse width is 1 μs resulting in a 150 m ambiguity in range to the scattering medium. As the overlap function can have large changes over this range, applying an accurate overlap correction becomes difficult, particularly at lower ranges where the overlap function changes most rapidly as a function of range.

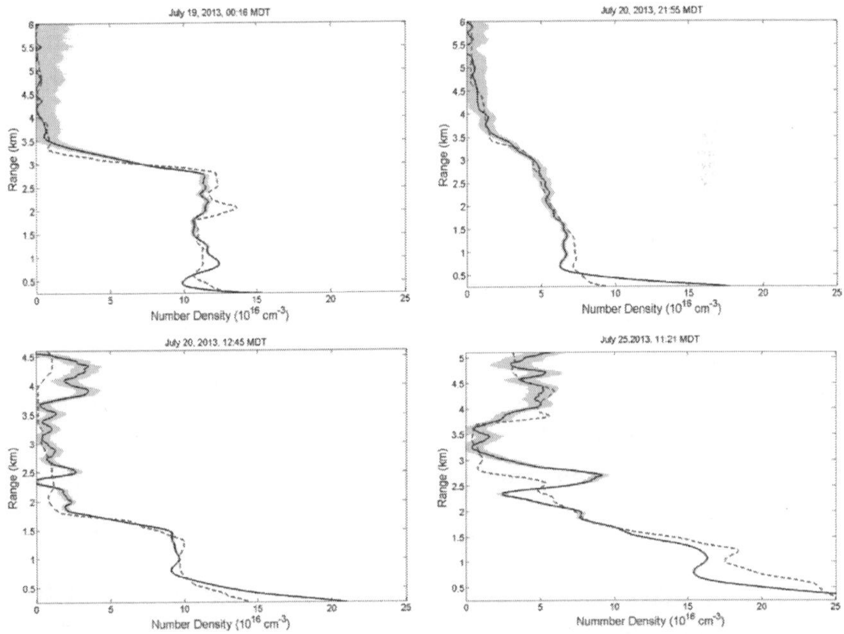

Figure 7. Plots of the water vapor number density as a function of range for data collect 19 July 2013 at 00:16, 20 July 2013 at 21:55, 20 July 2013 at 12:45, and 25 July 2013 at 11:21, are shown in the **top left**, **top right**, **bottom left**, and **bottom right** plots, respectively. The dashed red line represents retrieved number density profiles based on the temperature and relative humidity measured by the radiosonde while the solid black line represents the water vapor number density retrieved from the DIAL at the time of the radiosonde launch. The upper left-hand, upper right-hand, lower left-hand, and lower right-hand plots.

During daytime operation of the DIAL, background solar radiation will affect the signal to noise ratio of the DIAL instrument. A plot of the normalized relative backscatter as a function of range and time is shown in the upper plot inFigure 7 for 20 July 2013. At 00:00 MDT, aerosols are present to approximately 4 km. As the day progresses, the aerosol layer falls to below 2 km by 15:00 MDT then begins to rise again to approximately 4 km by 24:00 MDT. This pattern was seen on the clear sky days between 19 July 2013 and 23 July 2013. The water vapor number density as a function of range and time is shown in the middle plot of Figure 8. The water vapor distribution follows a similar pattern as the aerosol distribution. The bottom plot in Figure 8 shows the background count rate as a function of time. This count rate was determined by considering the count rate in the last 50 range bins used for background subtraction. On clear sky days, the background count rate remains below 2 MHz, well below the maximum count rate of the APD. At this count rate, using the manufacturer supplied data to determine a dead time of t_d = 39.2 ns, the correction factor reaches a value of 1.09. Water vapor profiles can be obtained up to 6 km during nighttime operation and between 4 and 6 km during daytime operation depending on atmospheric conditions.

A plot of the normalized relative backscatter is shown as a function of range and time as the upper plot in Figure 9 for 24 July 2013. Clouds where present between 2 km and 4 km between 05:00 MDT and 07:00 MDT and between 6 km and 10 km between 00:00 MDT and 02:00 MDT, 05:00 MDT and 12:00 MDT, and 14:00 MDT and 22:00 MDT. A plot of the water vapor number density as a function of range and time is shown in middle plot in Figure 9, while the background count rate as a function of time is shown in the bottom plot in Figure 9. The background count rate reached a maximum value of 14 MHz resulting in a correction factor of 2.2. When the background count rate begins to get large, the dead time of the APD starts to play a significant role in limiting the detection of signal photons at both the on-line and off-line wavelengths. This becomes particularly important at higher ranges where, over a one second averaging time, the signal is on the order of tens of photons. These missed photons due to the detector dead time have an impact on the water vapor retrieval and result in limiting the water vapor retrieval to approximately 3.5 km.

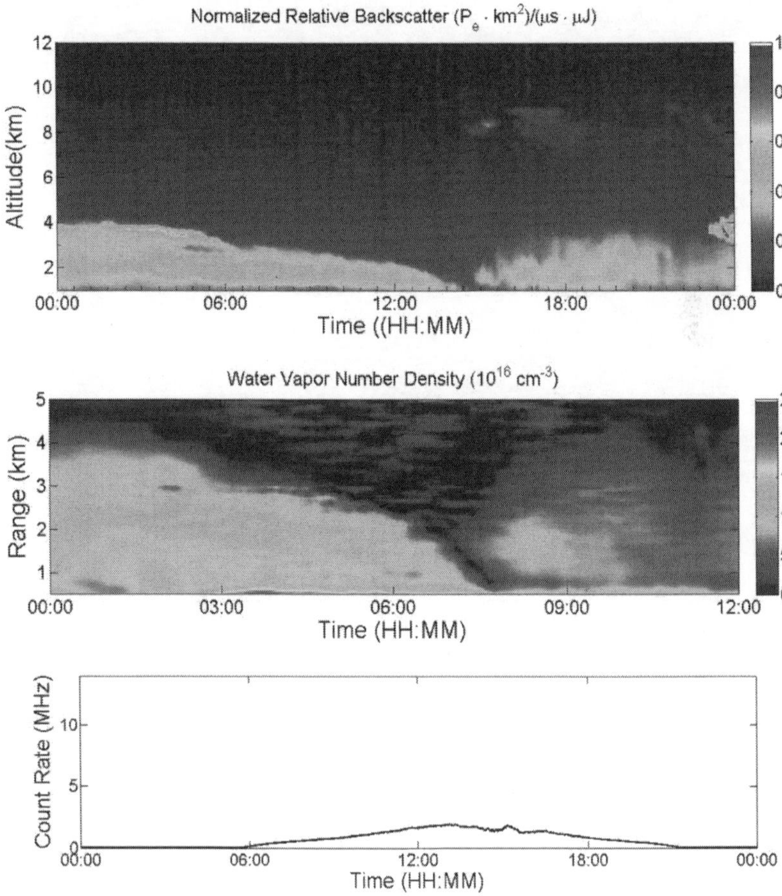

Figure 8. The upper plot shows the normalized relative backscatter as a function of time and range for 20 July 2013, while the middle figure shows the corresponding water vapor number density as a function of range and time. The bottom figure shows the background count rate as a function of time measured using the last fifty ranges bins. For clear sky operations, the maximum background count rate is less than 2 MHz and water vapor retrievals during daytime operations can be achieved between 4 and 6 km depending on atmospheric conditions.

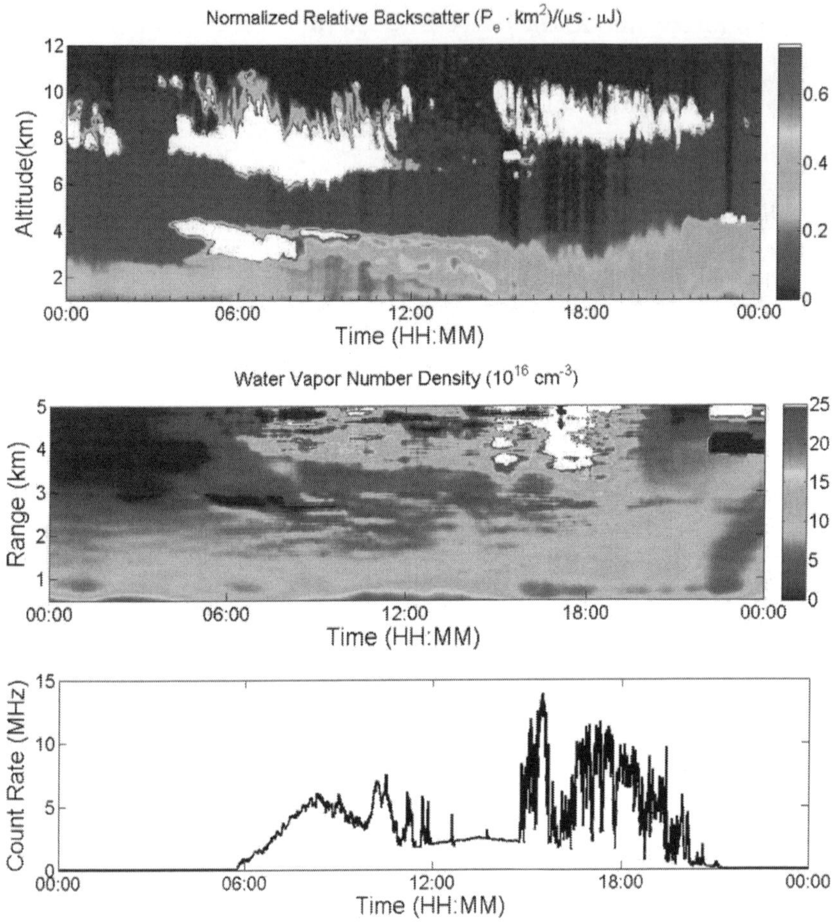

Figure 9. The upper plot shows the normalized relative backscatter as a function of time and range for 24 July 2013, while the middle figure shows the corresponding water vapor number density as a function of range and time. The bottom figure shows the background count rate as a function of time measured using the last fifty ranges bins. Operations during cloudy conditions result in maximum background count rates exceeding 14 MHz and water vapor retrievals during daytime operations can be achieved between 3 and 3.5 km depending on atmospheric conditions.

The improvements to the laser transmitter described in this paper are aimed at moving the DIAL from a research tool to a field deployable instrument, capable of autonomous operation over extended periods of time. One of the major drawbacks to the prototype DIAL instrument was the tendency of the external cavity diode lasers to mode hop causing the seed lasers to lose their frequency stability requiring monitoring of the DIAL during data collection. The incorporation of the DFB lasers into the laser transmitter has eliminated mode hoping allow long term data collection as demonstrated by the data presented in this section. Furthermore, the DFB laser allows continuous tuning over approximately 1 nm, compared to the approximately 0.02 nm of mode hop free tuning associated with the external cavity diode lasers. This ability to continuously tune the seed laser along the absorption feature is useful in adjusting the absorption cross section for prevailing meteorological conditions, which will become important for deploying the DIAL instrument at various field sites.

The DIAL laser transmitter described in this paper is an important step in developing an autonomous, field deployable DIAL for water vapor profiling in the lower troposphere. To complete this development, improvements to the DIAL receiver need to be incorporated. The DIAL instrument described in this paper utilized a 35.6 cm diameter Schmidt-Cassegrain telescope that achieves full overlap at 2.1 km. A two channel receiver configuration that includes a near field channel and a far field channel is currently being incorporated into the DIAL receiver. The far field channel will have the same configuration described in this paper while the near field channel will utilize an APD with free space coupling to achieve full overlap at less than 0.5 km. This configuration will utilize a second channel on the multichannel scaler card currently used in the DIAL instrument. The other major improvement to the DIAL receiver will involve the incorporation an etalon into the receiver with a free spectral range equal to the frequency spacing between the on-line and off-line wavelengths to block solar background light that passes through the current narrowband filters. This etalon has the potential to improve daytime performance, particularly during cloudy conditions by limiting the background count rate.

4. CONCLUSIONS

Water vapor is the most dominant greenhouse gas in the atmosphere and plays an important role in many key atmospheric processes associated with both weather and climate. Water vapor is highly variable in space and time due to large-scale transport and biosphere-atmosphere interactions. Having long-term, high-resolution, vertical profiles of water vapor will help to better understand the water vapor structure and variability and its associated impact on weather and climate. The laser transmitter based on tunable DBR lasers and a TSOA described in this paper is an important step in the realization of an autonomous DIAL instrument for network deploy for water vapor studies in the lower troposphere. This laser transmitter produced a pulsed output with a nominal pulse energy of 10 μJ, a pulse duration of 1 μs, and a pulse repetition frequency of 10 kHz. This laser transmitter was incorporated into a micro-pulse DIAL instrument and operated continuously over a 10-day period demonstrating its ability for continuous and autonomous operation.

ACKNOWLEDGMENTS

This work was supported under the kind auspices of the National Science Foundation grant number 1206166.

CONFLICTS OF INTEREST

The authors declare no conflict of interest.

REFERENCES

1. Emanuel, K.; Raymond, D.; Betts, A.; Bosart, L.; Bretherton, C.; Droegemeir, K.; Farrell, B.; Fritsch, J.M.; Houze, R.; Lemone, M.A.; *et al.* Report of the First Prospectus Development Team of the U.S. Weather Research Program to NOAA and the NSF. *Bull. Am. Meteor. Soc* **1995**, *76*, 1194–1208.

2. Dabberdt, W.F.; Schlatter, T.W. Research opportunities from emerging atmospheric observing and modeling capabilities. *Bull. Am. Meteor. Soc* **1996**, *77*, 305–323.

3. Weckwerth, T.M.; Wulfmeyer, V.; Wakimoto, R.M.; Hardesty, R.M.; Wilson, J.W.; Banta, R.M. NCAR-NOAA lower tropospheric water vapor workshop. *Bull. Am. Meteor. Soc* **1999**, *80*, 2339–2357.

4. Carbone, R.E.; Block, J.; Boselly, S.E.; Carmichael, G.R.; Carr, F.H.; Chandrasekar, V.; Gruntfest, E.; Hoff, R.H.; Krajewski, W.F.; LeMone, M.A.; *et al. Observing Weather and Climate from the Ground Up: A Nationwide Network of Networks*; National Academy Press: Washington, DC, USA, 2009.

5. Hooff, R.M.; Hardesty, R.M.; Carr, F.; Weckwerth, T.; Koch, S.; Benedetti, A.; Crewell, S.; Cimini, N.; Turner, D.; Feltz, W.; *et al. Thermodynamic Profiling Technologies Workshop, Report to the National Science Foundation and the National Weather Service*; Hardesty, R.M., Hoff, R.M., Eds.; National Center for Atmospheric Research: Boulder, CO, USA, 2011.

6. Whiteman, D.N.; Melfi, S.H.; Ferrare, R.A. Raman lidar system for the measurement of water vapor and aerosols in the Earth's atmosphere. *Appl. Opt* **1992**, *31*, 3068–3082.

7. Whiteman, D.N.; Rush, K; Rabenhorst, S.; Welch, W; Cadirola, M; McIntire, G; Russo, F.; Adam, M; Venable, D.; Connell, R.; *et al.* Airborne and ground-based measurements using a high-performance raman lidar. *J. Atmos. Oceanic Technol* **2010**, *27*, 1781–1801.

8. Wandinger, U. Raman Lidar. In *Lidar: Range-Resolved Optical Remote Sensing of the Atmosphere*; Weitkamp, C., Ed.; Springer: Berlin, Germany, 2005; pp. 241–271.

9. Whiteman, D. Examination of the traditional Raman lidar technique. II: Evaluating the ratios for water vapor and aerosols. *Appl. Opt* **2003**, *42*, 2593–2608.

10. Mattis, I.; Jaenisch, V. Automated Lidar Data Analyzer (ALDA) for RAMSES—The Autonomously Operating German Meteorological Service Raman Lidar for Atmospheric Moisture Sensing. Proceedings of 23rd International Laser Radar Conference, Nara, Japan, 24–28 July 2006; pp. 215–218.

11. Leblanc, T.; McDermid, I.S.; Aspey, R.A. First-year operation of a new water vapor Raman lidar at the JPL Table Mountain Facility, California. *J. Atmos. Ocean. Technol* **2008**, *25*, 1454–1462.

12. Higdon, N.S.; Browell, E.V.; Ponsardin, P.; Grossmann, B.E.; Butler, C.F.; Chyba, T.H.; Mayo, M.N.; Allen, R.J.; Heuser, A.W.; Grant, W.B.; *et al.* Airborne differential absorption lidar system for measurements of atmospheric water vapor and aerosols. *Appl. Opt* **1994**, *33*, 6422–6438.

13. Didier, B.; Quaglia, P.; Flamant, C.; Meissonnier, M.; Pelon, J. Airborne Lidar LEANDRE II for Water-Vapor Profiling in the Troposphere. I. System description. *Appl. Opt* **2001**, *40*, 3450–3461.

14. Wirth, M.; Fix, A.; Mahnke, P.; Schwarzer, H.; Schrandt, F.; Ehret, G. The airborne multi-wavelength water vapor differential

absorption lidar WALES: System design and performance. *Appl. Phys. B* **2009**, *96*, 201–213.

15. Machol, J.L.; Ayers, T.; Schwenz, K.T.; Koenig, K.W.; Hardesty, R.M.; Senff, C.J.; Krainak, M.A.; Abshire, J.B.; Bravo, H.E.; Sandberg, S.P. Preliminary measurements with an automated compact differential absorption lidar for profiling water vapor. *Appl. Opt* **2004**, *43*, 3110–3121.

16. Nehrir, A.R.; Repasky, K.S.; Carlsten, J.L.; Obland, M.D.; Shaw, J.A. Water vapor profiling using a widely tunable, amplified diode Laser based differential absorption lidar (DIAL). *J. Atmos. Ocean. Technol* **2009**, *26*, 733–745.

17. Nehrir, A.R. Water Vapor Profiling using a Compact Widely Tunable Diode Laser Differential Absorption Lidar (DIAL). M.S. Thesis. Montana State University, Bozeman, MT, USA, November 2008.

18. Nehrir, A.R.; Repasky, K.S.; Carlsten, J.L. Eye-safe diode-laser-based micropulse differential absorption lidar (DIAL) for water vapor profiling in the lower troposphere. *J. Atmos. Ocean. Technol* **2011**, *28*, 131–147.

19. Nehrir, A.R. Development of an eye-safe diode-laser-based micro-pulse differential absorption lidar (MP_DIAL) for atmospheric water vapor and aerosol studies, Ph.D. Thesis. Montana State University, Bozeman, MT, USA, July 2011.

20. Nehrir, A.R.; Repasky, K.S.; Carlsten, J.L. Micropulse water-vapor differential absorption lidar: transmitter design and performance. *Opt. Express* **2012**, *20*, 25137–25151.

21. Reagan, J.A.; Cooley, T.W.; Shaw, J.A. Prospects for an Economical, Eye Safe Water Vapor Lidar. Proceedings of the IEEE International Geoscience and Remote Sensing Symposium, Tokyo, Japan, 18–21 August, 1993; pp. 872–874.

22. Reagan, J.A.; Liu, H; McCalmont, J.F. Laser Diode Based New Generation Lidars. Proceedings of the IEEE International Geoscience and Remote Sensing Symposium, Lincoln, NE, USA, 27–31 May 1996; pp. 1535–1537.

23. Rothman, L.S.; Gordon, I.E.; Barbe, A.; Benner, D.C.; Bernath, P.F.; Birk, M.; Boudon, V.; Brown, L.R.; Campargue, A.; Champion, J.P.; *et al.* The HITRAN 2008 molecular spectroscopic database. *J. Quant. Spectrosc. Radiat. Transf* **2009**, *110*, 533–572.

24. Remsberg, E.E.; Gordley, L.L. Analysis of differential absorption lidar from the Space Shuttle. *Appl. Opt* **1978**, *17*, 624–630.

25. Wulfmeyer, V.; Bosenberg, J. Ground-based differential absorption lidar for water vapor profiling: Assesement of accuracy, resolution, and meteorological applications. *Appl. Opt* **1998**, *37*, 3825–3844.

26. Bosenberg, J. Ground-based differential absorption lidar for water-vapor profiling: Methodology. *Appl. Opt* **1998**, *37*, 3845–3860.

27. Siegman, A.A. *Lasers*; University Science Books: Sausalito, CA, USA, 1986.

28. Yariv, A. *Quantum Electronics*, 2nd ed.; John Wiley and Sons: New York, NY, USA, 1975.

29. Saleh, B.E.A.; Teich, M.C. *Fundamentals of Photonics*, 2nd ed; John Wiley and Sons: Hoboken, NJ, USA, 2007.

30. Wulfmeyer, V.; Walther, C. Future performance of ground-based and airborne water-vapor differential absorption lidar. I. overview and theory. *Appl. Opt* 2001, *40*, 5304–5320.

4

A HIGH-PERFORMANCE QUANTUM DOT SUPERLUMINESCENT DIODE WITH A TWO-SECTION STRUCTURE

Xinkun Li, Peng Jin, Qi An, Zuocai Wang, Xueqin Lv, Heng Wei, Jian Wu, Ju Wu and Zhanguo Wang

Key Laboratory of Semiconductor Materials Science, Institute of Semiconductors, Chinese Academy of Sciences, Beijing, 100083, China

ABSTRACT

Based on InAs/GaAs quantum dots [QDs], a high-power and broadband superluminescent diode [SLD] is achieved by monolithically integrating a conventional SLD with a semiconductor optical amplifier. The two-section QD-SLD device exhibits a high output power above 500 mW with a broad emission spectrum of 86 nm. By properly controlling the current injection in the two sections of the QD-SLD device, the output power of the SLD can be tuned over a wide range from 200 to 500 mW while preserving a broad emission spectrum based on the balance between the ground state emission and the first excited state emission of QDs. The gain process of the two-section QD-SLD with different pumping levels in the two sections is investigated.

KEYWORDS

quantum dot, superluminescent diode, two-section structure, optical amplification

1. INTRODUCTION

Superluminescent diodes [SLDs] have attracted extensive attention for a wide range of applications, such as optical coherence tomography [OCT] [1, 2], optical fiber-based sensors [3–5], external cavity tunable lasers [6–8], optoelectronic systems [9], etc. A wide emission spectrum corresponding to a low degree of coherence is required for these applications of SLD, which allows the realization of sensors with improved resolution. It has been proposed that self-assembled quantum dots [QDs] [10–12] and quantum well grown on a high-index surface are beneficial to broaden the spectral bandwidth of the device [13]. Till now, QDs have successfully been used as the active media in several broadband light-emitting devices, such as QD-SLDs [14–20], QD semiconductor optical amplifiers [SOAs] [21–23], and QD broadband laser diodes [24–26]. For QD-SLD devices, a high power of 200 mW [14] and a wide spectral bandwidth of more than 140 nm [27, 28] have been achieved. Most recently, an intermixed QD-SLD exhibits a power of 190 mW with a 78-nm spectral bandwidth [29].

For a typical SLD device structure with a single current-injection section, the high output power can only be obtained at a high pumping level, where the device demonstrates a narrow spectrum emitted predominantly from the QDs' excited state [ES] due to the low saturated gain of the QD ground state [GS]. It is difficult to achieve high-power and broad-emitting spectrum simultaneity. However, a high-power SLD that is broadband emitting is required in some fields. As an example, in an OCT system, a high power is usually needed to enable greater penetration depth and improve the imaging sensitivity [30]. Numerical investigation [31] and experimental evidence [32, 33] have shown that this limitation can be overcome by using a multi-section structure in an SLD device, which allows the emission spectrum and output power to be tuned independently. A quantum-well SLD with a two-section structure which integrates monolithically an SLD with an SOA has been reported, which exhibits an output power that is one or two orders of magnitude higher than that in conventional SLD devices [34].

In this paper, a QD-SLD device, which has a two-section structure monolithically integrating an SLD with an SOA, is fabricated. A high power (500 mW) with a broad emission of 86 nm is obtained. By properly controlling the current injection in the two sections of the QD-SLD device, the power tunability over a wide range from 200 to 500 mW is achieved, with the preservation of a nearly constant spectral width.

2. EXPERIMENT

The epitaxial structure of the QD-SLD device in this study was grown by a Riber 32P solid-source molecular beam epitaxy machine on n-GaAs(001) substrate. The epitaxial structure consists of ten InAs-QD layers separated from each other by a GaAs spacer; each of them is formed by depositing a 1.8-monolayer InAs at 480°C and covered by a 2-nm $In_{0.15}Ga_{0.85}As$. Ten QD layers plus the GaAs waveguide layers form the whole active region which is sandwiched between 1.5-μm n- and p-type $Al_{0.5}Ga_{0.5}As$ cladding layers. Finally, a p^+-doped GaAs contact layer completes the structure.

A QD-SLD device with an index-guided ridge waveguide and a two-section structure was fabricated. A schematic diagram of the geometrical design (not to scale) is shown in Figure 1. The device integrates monolithically an SLD with a tapered SOA. The SLD section is 1-mm long and 10-μm wide. The tapered SOA section is 3-mm long with a full flare angle of 6°. The ridge waveguide was fabricated using photolithography and wet chemical etching. The center axis of the ridge is aligned at 6° with respect to the facet normal to suppress lasing. A 200-μm-length output window structure (no electric contact) is used to reduce the risk for catastrophic optical damage of the output facet with a high output power. Ti/Au and AuGeNi/Au ohmic contacts were evaporated on the top and back of the wafer, respectively. A 20-μm-wide separation between the SLD and the SOA sections is realized by removing the upper Ti/Au ohmic contact and the 0.5-μm epilayer using photolithography and wet chemical etching. After metallization, the device was cleaved and mounted p-side up on a copper sink using an indium solder. Antireflection coatings of λ/4 were used on both facets of the device. The QD-SLD device was characterized by light power-injection current [*P-I*] and electroluminescence measurements at room temperature under a pulsing (1 kHz repetition rate and 3% duty cycle) injection in the SOA section and a continuous-wave injection in the SLD section, respectively.

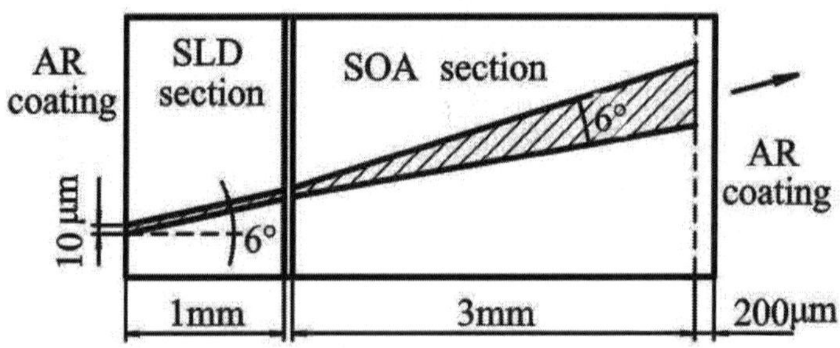

Figure 1 Schematic diagram of the QD-SLD device with a two-section structure.

3. RESULTS AND DISCUSSION

Figure 2 shows the *P-I* characteristic of the SOA section with the SLD section un-pumped and acting as a rear optical absorption region. A superluminescent characteristic is clearly observed by the superlinear increase in optical power with the current. At a current of 9.8 A, a maximum output power of 280 mW is obtained. The emission spectra under different injection currents in the SOA section [I_{SOA}] are shown in the inset of Figure 2. When I_{SOA} = 2 A, the center wavelength of the emission spectrum is 1.18 μm with a full width at half maximum of 43 nm, which corresponds to the QDs' GS emission. The relatively wide GS emission is attributed to the size inhomogeneity that is naturally occurring in self-assembled QDs. With the increasing I_{SOA}, the emission spectra are clearly broadened to the short-wavelength side, which should be attributed to the sequential carrier filling into the first ES [ES1]. For a given I_{SOA} of 8.35 A, due to the nearly identical contribution to the emission from the QDs' GS and ES1, a 94-nm broad spectrum with a power of 200 mW is achieved.

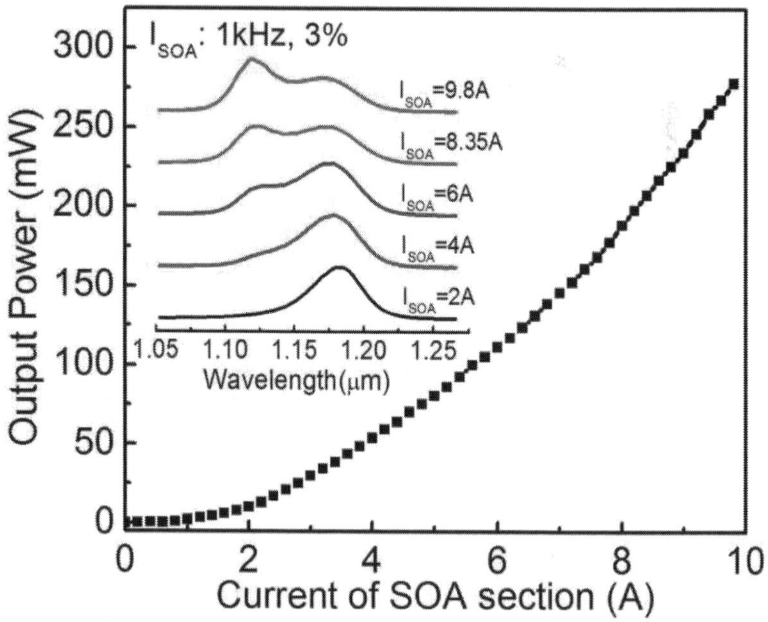

Figure 2 ***P - I*** **characteristic of the SOA section with the SLD section un-pumped**. The inset shows the normalized emission spectra under different injection currents of the SOA section.

The characteristics of the two-section SLD device were measured when the SLD section was pumped to seed the SOA section. The output-power characteristics versus I_{SOA} under different SLD section currents [I_{SLD}] are shown in Figure 3. It can be seen from the figure that the output power increases rapidly with the increasing current injection in the SLD section. Without pumping the SLD section, the output power of the device is 280 mW at I_{SOA} = 9.8 A. The output power can reach 1.15 W at I_{SOA} = 9.8 A and I_{SLD} = 400 mA. The device begins lasing when the power is in the range of 500 to approximately 600 mW with various SOA and SLD current combinations (refer to Figure 4). The evident increase of output power is attributed to the amplification of the input beam while propagating forward from the narrow end to the wide end of the tapered region. With a full flare angle of 6°, the incident beam will expand freely to fill the full tapered region owing to diffraction [35]. The optical density will be reduced, which increases the saturated power.

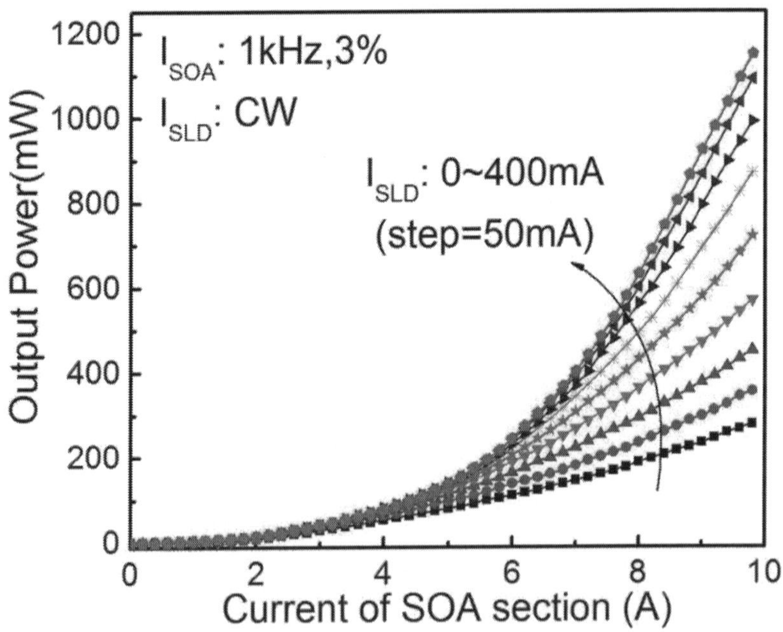

Figure 3 Output power versus SOA current under different injection currents of the SLD section.

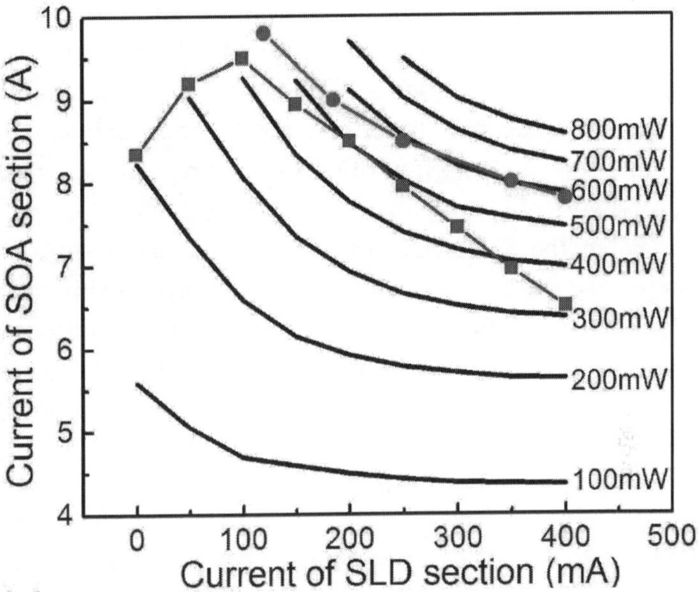

Figure 4 Equal power curves (solid lines) as function of the currents injected in the two sections. The solid squares show the combinations of currents at which QDs' GS and 1st ES give equivalent contributions to the emission spectra. Current combinations at which the device begins lasing are shown in solid circles.

The emission spectra measured from the SOA facet under different I_{SLD}, with I_{SOA} fixed at 6.5, 8.5, and 9.5 A, respectively, are shown in Figure 5. As expected, it can be seen from the figure that the spectrum shape and emission bandwidth of the QD-SLD device with the two-section structure can be tuned by properly controlling the current injection in the two sections. With I_{SOA} fixed at 6.5 A as shown in Figure 5a, the GS emission provides the main contribution to the spectrum when the SLD section is not pumped. To obtain a more broadened emission bandwidth based on the balance of the QDs' GS and ES1 emissions, the input beam from the SLD section must provide a greater amount of ES emission. When the SLD section is driven with 400 mA of current-injection in order to seed the SOA section, the resultant emission of the QDs' GS and ES1 have nearly equivalent contributions, and a 76-nm bandwidth is obtained. At this working point, the device gives a 320-mW power output. Similarly, a broad emission spectrum based on the balance between the GS emission and the ES1 emission of QDs is achieved at I_{SLD} = 200 and 100 mA for a given I_{SOA} of 8.5 and 9.5 A, respectively. With I_{SOA} fixed at 8.5 A, when the SLD section is driven with a 200-mA current-injection to seed the SOA section, the QD-SLD device exhibits a broad emission spectrum of 86 nm and a simultaneous high output power of 504 mW. For a given I_{SOA} of 9.5 A, with the SLD section un-pumped, the ES1 emission provides the main contribution to the emission spectrum. In order to achieve a balanced emission from GS and ES1, the GS-dominated emission is introduced to the SOA using I_{SLD} = 100 mA. As a result, the resultant contribution of the QDs' GS and ES1 is equivalent. A broad emission spectrum of 88 nm with the output power of 422 mW is obtained.

It can be seen from the above results that the output power and spectrum bandwidth can be tuned by properly controlling the current densities injected in the two regions of the QD-SLD. Figure 4 shows equal power curves as function of the currents injected in the two sections. Data points (solid squares) at which the GS and ES1 have nearly identical emission intensities, corresponding to the maximum bandwidth of the emission spectrum, are also shown in Figure 4. It can be found that the output power can be tuned over a wide range of 200 to 500 mW while preserving a broad emission spectrum. The high output power and wide power tunability is due to the two-section structure which integrates a tapered SOA section. Current combinations at which the device begins lasing are also shown in Figure 4 (solid circles). Working points of the QD-SLD device can be set in the lower left region of the borderline. An optimum working point is found in the figure that the SOA current is in the 8- to approximately 8.5-A range and the SLD current is 0.2 to approximately 0.25 A, at which a 500-mW output power and an 86-nm bandwidth are achieved simultaneously.

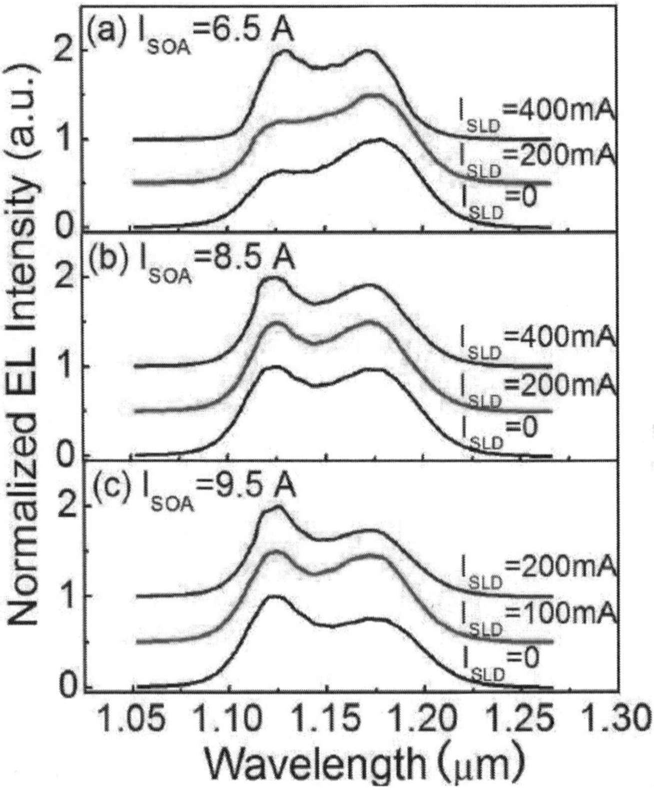

Figure 5 Normalized emission spectra from the SOA facet under different pumps of the SLD section. They are for a given SOA injection of (**a**) 6.5, (**b**) 8.5, and (**c**) 9.5 A, respectively. Some spectra are shifted vertically for clarity.

4. CONCLUSION

In conclusion, a high-power QD SLD with a broad bandwidth in the emission spectra is achieved by the two-section structure which monolithically integrates an SLD with a tapered SOA. Properly controlling the current densities injected in the two sections, the QD-SLD device exhibits a maximum output power above 500 mW and a simultaneously broad bandwidth of 86 nm. Also, the output power can be tuned over a wide range from 200 to 500 mW while preserving a nearly constant spectral width.

ACKNOWLEDGEMENTS

This work was supported by the National Basic Research Program of China (no. 2006CB604904) and the National Natural Science Foundation of China (nos. 60976057, 60876086, and 60776037).

COMPETING INTERESTS

The authors declare that they have no competing interests.

AUTHORS CONTRIBUTIONS

XL carried out the device process, device characterization, and data analysis; participated in the experimental design; and drafted the manuscript. PJ conceived the study, participated in its design and coordination, and performed the epitaxial growth. QA participated in the data analysis. ZW participated in its design and carried out some preparative work. XL participated in the epitaxial growth. HW participated in the device process. JW participated in the device process. JW modified the draft. ZW conceived the study. All authors read and approved the final manuscript.

REFERENCES

1. Schmitt JM: Optical coherence tomography (OCT): a review. *IEEE J Sel Topics Quantum Electron* 1999, 5: 1205–1215.

2. Zotter S, Pircher M, Torzicky T, Bonesi M, Götzinger E, Leitgeb RA, Hitzenberger CK: Visualization of microvasculature by dual-beam phase-resolved Doppler optical coherence tomography. *Opt Express* 2011, 19: 1217–1227.

3. Burns WK, Chen C, Moeller RP: Fiber-optic gyroscopes with broad-band sources. *J Lightw Technol* 1983, 1: 98–105.

4. Lee B: Review of the present status of optical fiber sensors. *Opt Fiber Technol* 2003, 9: 57–79.

5. Krstajić N, Childs D, Smallwood R, Hogg R, Matcher SJ: Common path Michelson interferometer based on multiple reflections within the sample arm: sensor applications and imaging artifacts. *Meas Sci Technol* 2011, 22: 027002.

6. Lv XQ, Jin P, Wang ZG: Broadly tunable grating-coupled external cavity laser with quantum-dot active region. *IEEE Photon Technol Lett*2010, 22: 1799–1801.

7. Lv XQ, Jin P, Wang WY, Wang ZG: Broadband external tunable quantum dot lasers with low injection current density. *Opt Express* 2010, 18: 8916–8922.

8. Fedorova KA, Cataluna MA, Krestnikov I, Livshits D, Rafailov EU: Broadly tunable high-power InAs/GaAs quantum-dot external cavity diode lasers. *Opt Express* 2010, 18: 19438–19443.

9. Li X, Cohen AB, Murphy TE, Roy R: Scalable parallel physical random number generator based on a superluminescent LED. *Opt Lett*2011, 36: 1020–1022.

10. Sun Z-Z, Ding D, Gong Q, Zhou W, Xu B, Wang ZG: Quantum-dot superluminescent diode: a proposal for an ultra-wide output spectrum.*Opt Quantum Electron* 1999, 31: 1235–1246.

11. Meng XQ, Jin P, Liang ZM, Liu FQ, Wang ZG, Zhang ZY: Structure and properties of InAs/AlAs quantum dots for broadband emission.*J Appl Phys* 2010, 108: 103515.

12. Wu J, Wang ZM, Dorogan VG, Li S, Mazur YI, Salamo GJ: Near infrared broadband emission of $In_{0.35}Ga_{0.65}As$ quantum dots on high index GaAs surfaces. *Nanoscale* 2011, 3: 1485–1488.

13. Li Z, Wu J, Wang ZM, Fan D, Guo A, Li S, Yu S, Manasreh O, Salamo GJ: InGaAs quantum well grown on high-index surfaces for superluminescent diode applications. *Nanoscale Res Lett* 2010, 5: 1079–1084.

14. Zhang ZY, Wang ZG, Xu B, Jin P, Sun ZZ, Liu FQ: High-performance quantum-dot superluminescent diodes. *IEEE Photon Technol Lett*2004, 16: 27–29.

15. Li LH, Rossetti M, Fiore A, Occhi L, Velez C: Wide emission spectrum from superluminescent diodes with chirped quantum dot multilayers.*Electron Lett* 2005, 41: 41–43.

16. Rossetti M, Markus A, Fiore A, Occhi L, Velez C: Quantum dot superluminescent diodes emitting at 1.3 μm. *IEEE Photon Technol Lett*2005, 17: 540–542.

17. Djie HS, Dimas CE, Wang D-N, Ooi B-S, Hwang JCM, Dang GT, Chang WH: InGaAs/GaAs quantum-dot superluminescent diode for optical sensor and imaging. *IEEE Sensors J* 2007, 7: 251–257.

18. Zhang ZY, Luxmoore IJ, Jin CY, Liu HY, Jiang Q, Groom KM, Childs DT, Hopkinson M, Cullis AG, Hogg RA: Effect of facet angle on effective facet reflectivity and operating characteristics of quantum dot edge emitting lasers and superluminescent light-emitting diodes.*Appl Phys Lett* 2007, 91: 081112.

19. Haffouz S, Rodermans M, Barrios PJ, Lapointe J, Raymond S, Lu Z, Poitras D: Broadband superluminescent diodes with height-engineered InAs-GaAs quantum dots. *Electron Lett* 2010, 46: 1144–1146.

20. Zhang ZY, Jiang Q, Hopkinson M, Hogg RA: Effects of intermixing on modulation p-doped quantum dot superluminescent light emitting diodes. *Opt Express* 2010, 18: 7055–7063.

21. Bakonyi Z, Su H, Onishchukov G, Lester LF, Gray AL, Newell TC, Tünnermann A: High-gain quantum-dot semiconductor optical amplifier for 1300 nm. *IEEE J Quantum Electron* 2003, 39: 1409–141.

22. Wong HC, Ren GB, Rorison JM: Mode amplification in inhomogeneous QD semiconductor optical amplifiers. *Opt Quantum Electron* 2006, 38: 395–409.

23. Park J, Kim NJ, Jang YD, Lee EG, Lee JM, Baek JS, Kim JH, Lee HS, Yee KJ, Lee D, Pyun SH, Jeong WG, Kim J: Gain dynamics of an InAs/InGaAsP quantum dot semiconductor optical amplifier operating at 1.5 μm. *Appl Phys Lett* 2011, 98: 011107.

24. Sugawara M, Mukai K, Nakata Y: Light emission spectra of columnar-shaped self-assembled InGaAs/GaAs quantum-dot lasers: effect of homogeneous broadening of the optical gain on lasing characteristics. *Appl Phys Lett* 1999, 74: 1561–1563.

25. Kovsh A, Krestnikov I, Livshits D, Mikhrin S, Weimert J, Zhukov A: Quantum dot laser with 75 nm broad spectrum of emission. *Opt Lett*2007, 32: 793–795.

26. Zhukov AE, Kovsh AR: Quantum dot diode lasers for optical communication system. *Quantum Electron* 2008, 38: 409–423.

27. Zhang ZY, Hogg RA, Xu B, Jin P, Wang ZG: Realization of extremely broadband quantum-dot superluminescent light-

emitting diodes by rapid thermal-annealing process. *Opt Lett* 2008, 33: 1210–1212.

28. Lv XQ, Liu N, Jin P, Wang ZG: Broadband emitting superluminescent diodes with InAs quantum dots in AlGaAs matrix. *IEEE Photon Technol Lett* 2008, 20: 1742–1744.

29. Jiang Q, Zhang ZY, Hopkinson M, Hogg RA: High performance intermixed p-doped quantum dot superluminescent diodes at 1.2 μm.*Electron Lett* 2010, 46: 295-U49.

30. Brezinski ME, Fujimoto JG: Optical coherence tomography: high-resolution imaging in nontransparent tissue. *IEEE J Sel Topics Quantum Electron* 1999, 5: 1185–1192.

31. Rossetti M, Bardella P, Montrosset I: Numerical investigation of power tenability in two-section QD superluminescent diodes. *Opt Quantum Electron* 2008, 40: 1129–1134.

32. Xin Y-C, Martinez A, Saiz T, Moscho AJ, Li Y, Nilsen TA, Gray AL, Lester LF: 1.3-μm quantum-dot multisection superluminescent diodes with extremely broad bandwidth. *IEEE Photon Technol Lett* 2007, 19: 501–503.

33. Greenwood PDL, Childs DTD, Groom KM, Stevens BJ, Hopkinson M, Hogg RA: Tuning superluminescent diodes characteristics for optical coherence tomography systems by utilizing a multicontact device incorporating wavelength-modulated quantum dots. *IEEE J Sel Topics Quantum Electron* 2009, 15: 757–763.

34. Du GT, Devane G, Stair KA, Wu SL, Chang RPH, Zhao YS, Sun ZZ, Liu Y, Jiang XY, Han WH: The monolithic integration of a superluminescent diode with a power amplifier. *IEEE Photon Technol Lett* 1998, 10: 57–59.

35. Walpole JN: Semiconductor amplifiers and lasers with tapered gain regions. *Opt Quantum Electron* 1996, 28: 623–645.

5

FREQUENCY-CONVERTED DILUTE NITRIDE LASER DIODES FOR MOBILE DISPLAY APPLICATIONS

Janne Konttinen and Ville-Markus Korpijärvi

1EpiCrystals Oy, Hermiankatu 12 E 2, Tampere 33720, Finland.

2Optoelectronics Research Centre, Tampere University of Technology, Korkeakoulunkatu 3, Tampere 33720, Finland.

ABSTRACT

We demonstrate a 1240-nm GaInNAs multi-quantum well laser diode with an integrated saturable electro-absorber whose wavelength is converted to 620 nm. For conversion, we used a MgO:LN nonlinear waveguide crystal with an integrated Bragg grating in direct coupling configuration. Broadened visible spectral width and reduced speckle as well as a high extinction ratio between the below and above threshold powers were observed in passively triggered pulsed operation with smooth direct current modulation characteristics. The demonstration opens a new avenue for developing 620-nm semiconductor lasers required for emerging projection applications.

KEYWORDS

Semiconductor laser, Second harmonic generation, Visible III-V semiconductors, Dilute nitride, GaInNAs Mobile display

1. BACKGROUND

Red laser light sources emitting in the wavelength range of 610 to 620 nm are particularly interesting for mobile display applications due to increased luminous efficacy and higher achievable brightness within eye-safety regulations [1]. Unfortunately, this wavelength range is difficult to achieve by using traditional GaInP/AlGaInP red laser diodes (LDs) [2]. Another well-known drawback of GaInP/AlGaInP diodes is the reduction of characteristic temperature of threshold current (T_0) with wavelength. HighT_0 values have been demonstrated with red laser diodes emitting at wavelengths above 650 nm [3], while shorter wavelength diodes suffer from poor temperature characteristics [4]. These features render impossible the use of standard AlGaInP laser diodes in embedded projection displays, where large operating temperature range is typically required.

Frequency conversion of infrared laser emission is an attractive solution for the generation of short-wavelength red light [5]. While GaInAs quantum well (QW) emission wavelength is practically limited to approximately 1200 nm [6], by using dilute nitride GaInNAs QWs with a tiny fraction of nitrogen added to the highly strained GaInAs, the emission wavelength can be extended to 1220-1240 nm for high luminosity red light generation at 610 to 620 nm by frequency conversion [5]. In addition, excellent temperature characteristics and high power operation have been demonstrated with GaInNAs laser diodes in this wavelength range [7].

2. METHODS

The GaInNAs/GaAs semiconductor heterostructure was grown on an n-GaAs (100) substrate by Veeco (Plainview, NY, USA) GEN20 molecular beam epitaxy (MBE) reactor with a radio frequency plasma source for nitrogen, a valved cracker for arsenic, and normal effusion cells for the group-III materials and dopants. Silicon and beryllium were used as n- and p-type dopants. The active region of the laser structure consisted of two 7-nm thick GaInNAs QWs separated by a 20-nm GaAs layer. The $Ga_{1-x}In_xN_yAs_{1-y}$ QWs had the nominal indium and nitrogen compositions of $x = 33.6\%$ and $y = 0.6\%$, respectively. This double-QW structure was embedded in GaAs whose thickness was 142 nm on both sides of the structure. The undoped waveguide structure was surrounded by 1.5-μm thick n-$Al_{0.30}Ga_{0.70}As$ on the substrate side and 1.5 μm p-$Al_{0.30}Ga_{0.70}As$ on the top side. On top of the p-AlGaAs cladding, a p-GaAs contact layer was grown to finalize the structure. Figure 1 shows the band gap profile of the structure and summarizes the layer thicknesses. Strong room-temperature photoluminescence (PL) emission measured from this structure peaked at 1231 nm, as shown in Figure 2. Two heterostructures, comprising one or two QWs, were considered for the frequency-doubled 620-nm laser demonstration. The single-QW and double-QW structures were compared as broad-area ridge-waveguide (RWG) lasers in pulsed current mode. The double-QW structure was opted because it showed only slightly higher threshold current as compared with the single-QW structure (adding the second QW to the test structure increased the threshold current density from 500 to 610 A/cm^2), and double-QW lasers are known to be less temperature sensitive, i.e., to have largerT_0[8], which is important for the targeted application. The difference between the slope efficiency values of the single-QW and double-QW structures was negligible.

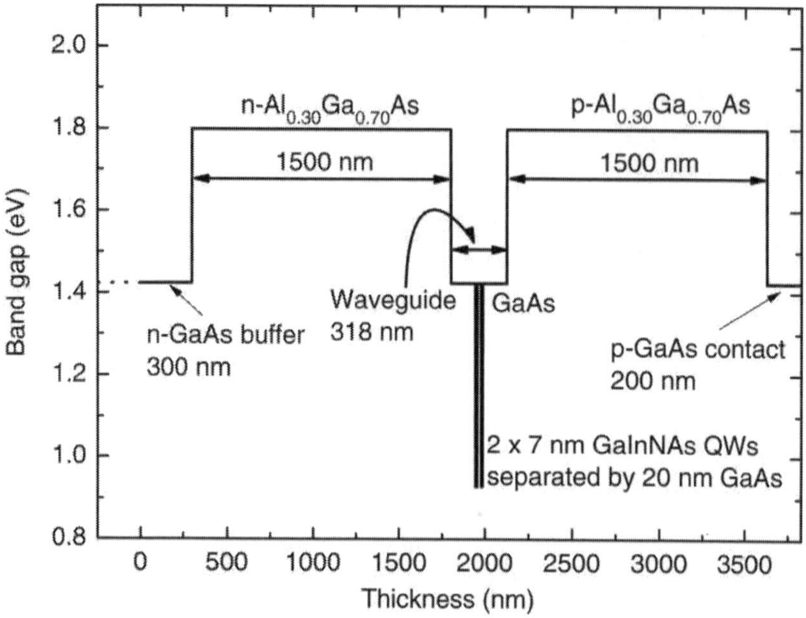

Figure 1 Band gap profile and layer thicknesses of the semiconductor heterostructure of the 1240-nm GaInNAs laser.

The processed laser chips employed a single transverse mode RWG process with ridge width of 3.5 μm and cavity length of 1250 μm. The laser diode further comprised an 85-μm reverse-biased saturable electro-absorber section to passively trigger short pulses for enhancing frequency conversion efficiency in the nonlinear waveguide. The front and rear facets of the laser diode were AR/HR coated with reflectivities of <1% and >95% at 1240 nm, respectively.

A nonlinear waveguide crystal made of MgO-doped $LiNbO_3$ with high nonlinear coefficient was used for frequency doubling to visible wavelengths. The crystal had a surface Bragg grating implemented near the output end of the waveguide. The function of the surface Bragg grating is to provide self-seeding to frequency lock the IR laser diode in order to maintain sufficient spectral overlap with acceptance spectrum of quasi-phase-matched (QPM) grating over an extended temperature range.

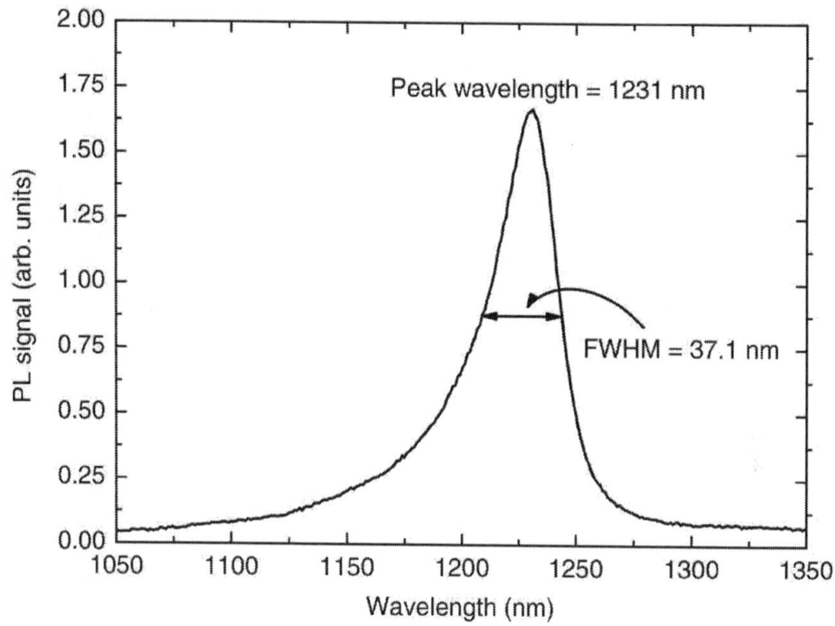

Figure 2 Room-temperature PL emission measured from the 1240-nm GaInNAs/GaAs laser wafer.

3. RESULTS AND DISCUSSION

3.1 Free-running performance

In free-running mode with the absorber section unbiased, the 1240-nm RWG laser diode exhibited an average slope efficiency of approximately 0.7 W/A and smooth *L-I* characteristics at 25°C as shown in Figure 3. The temperature performance was investigated in continuous wave (CW) mode (i.e. the absorber section forward biased by a contact to gain section). Kink-free operation up to 300 mA was demonstrated over the temperature range from 25°C to 60°C, as shown in Figure 4. The corresponding characteristic temperature (T_0) was 97 K for the low front-facet reflectivity-coated free-running LDs (see Figure 4). As it can be seen in Figure 5, the lateral far field exhibited stable single-mode operation up to 350 mA with no evidence of beam steering. The beam opening angles (FWHM) were 40° and 17° for fast and slow axes, respectively. Comparing the measured threshold current and T_0 values with the values of related red AlGaInP-based laser diodes is difficult, because these lasers can hardly reach lasing at 620 nm at normal temperature and pressure. Commercial single-transverse-mode RWG laser diode operating at longer wavelengths (633 nm) [9] has a threshold current of about 60 mA at 25°C, which is identical to the value of the GaInNAs laser reported here. Based on the data available on the datasheet [9], the T_0 of this commercial laser diode is estimated to be 89 K, which comes close to the value reported here for the GaInNAs laser. However, the T_0 value of free-running GaInNAs diode is suppressed due to the low front-facet reflectivity [10] and can thus be improved by providing the wavelength locking optical feedback from Bragg grating in nonlinear waveguide [11]. In addition, it is known that the performance of AlGaInP-based laser diodes, especially their T_0 values, deteriorate strongly as the wavelength is decreased towards 620 nm [4, 12, 13].

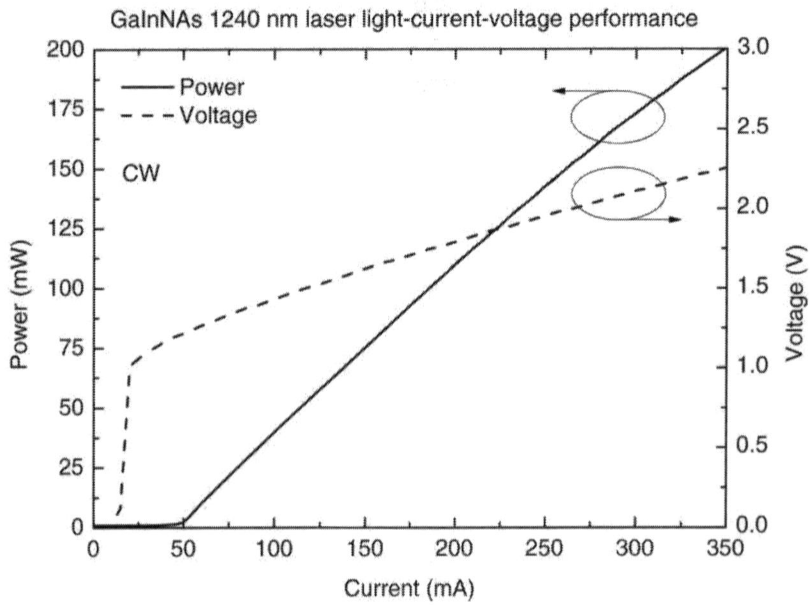

Figure 3 Continuous wave performance of a single-mode 1240-nm GaInNAs laser diode.

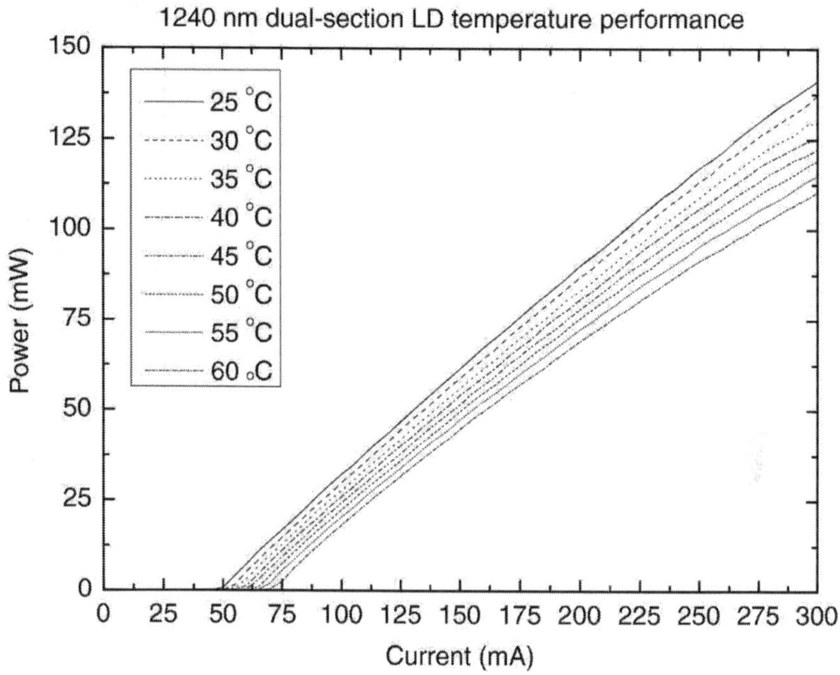

Figure 4 Continuous wave performance of a single-mode 1240-nm GaInNAs laser diode at elevated temperatures.

3.2 Frequency conversion

The passively pulsed frequency-converted 620-nm laser configuration is shown in Figure 6. The 1240-nm infrared emission from the GaInNAs laser diode is directly coupled to MgO:LN waveguide for single-pass frequency conversion. The surface Bragg grating is implemented near the output end of the nonlinear waveguide, while the reverse-biased saturable absorber is located near the highly reflective back facet of the laser diode. Both facets of the nonlinear waveguide, as well as the output facet of the laser diode, are AR-coated to suppress interface reflections.

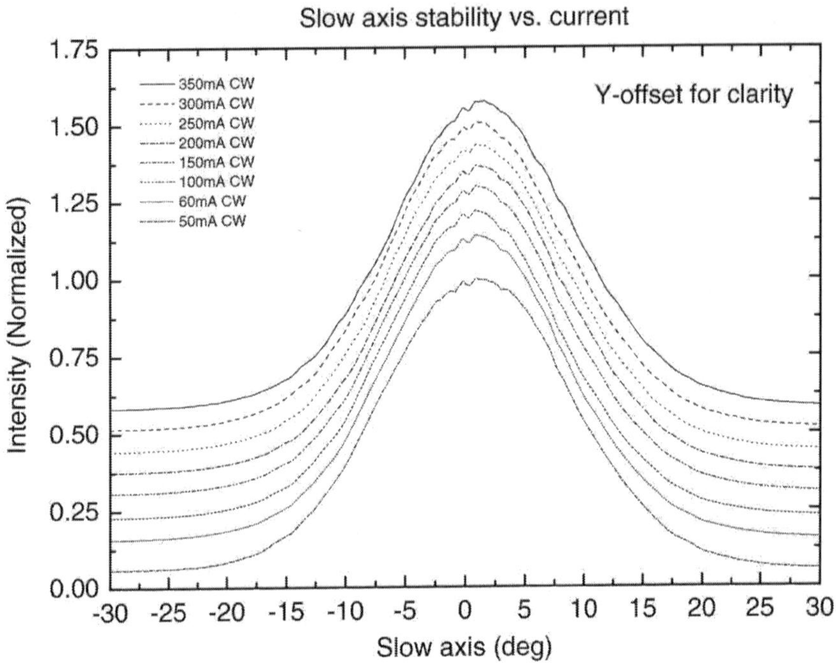

Figure 5 Lateral far-field stability vs. current in continuous wave mode at room temperature.

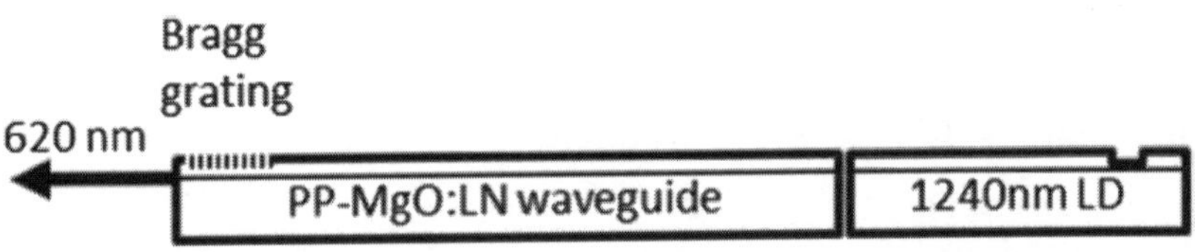

Figure 6 Coupling configuration of passively pulsed frequency-converted 620-nm laser.

Successful wavelength locking and passively pulsed operation (with absorber reverse biased) are achieved with the direct coupling configuration between the GaInNAs laser diode and MgO:LN waveguide. The infrared and visible spectra were recorded using

Yokogawa AQ6373 optical spectrum analyzer (Tokyo, Japan) with extended wavelength range. Compared with the CW mode, the infrared (Figure 7) and visible spectra (Figure 8) are broadened when the absorber section was biased with 0.4- to 1.5-V reverse-bias voltage triggering passively pulsed mode. A considerable reduction in the speckle visibility is observed under pulsed mode when compared with continuous wave operation. This observation is supported by the measured broadening of the visible spectrum.

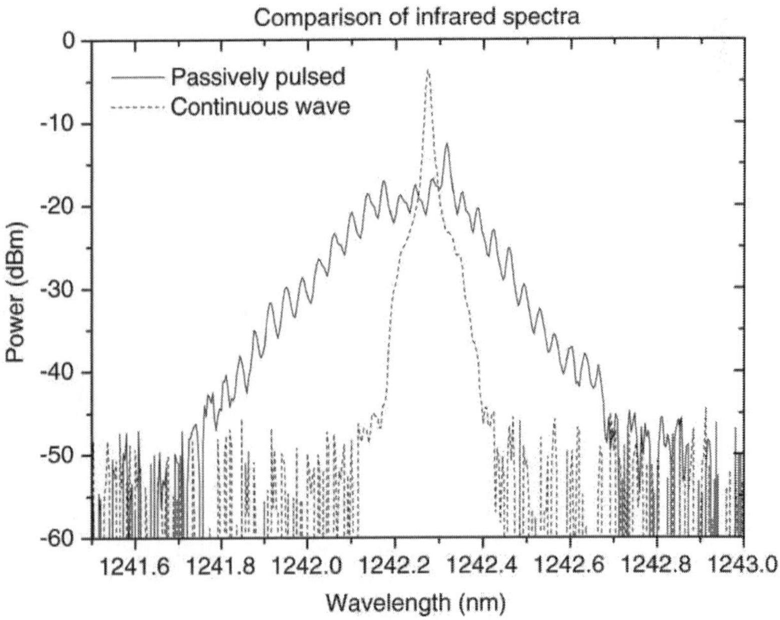

Figure 7 Comparison of grating-locked infrared spectra under continuous wave (dashed line) and pulsed (solid line) operating modes.

The *L-I-V* performance under the passively pulsed reverse-biased mode was investigated using 0.2-mA current resolution in the visible output power range of 0 to 1 mW, as targeted for near-to-eye display applications. The lasing threshold was 63 mA under 0.4-V reverse bias. Above the lasing threshold, the visible light output represented smooth, slightly non-linear *L-I* curve within the targeted operating power range. The results are summarized in Figure 9.

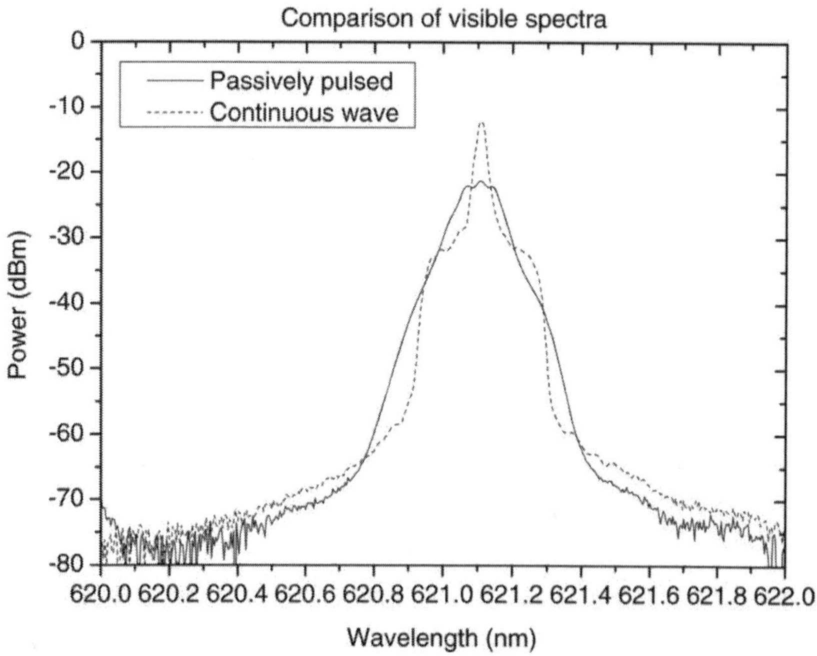

Figure 8 Comparison of grating-locked visible spectra under continuous wave (dashed line) and pulsed (solid line) operating modes.

The exceptional feature of the 620-nm frequency converted visible light source with 'no visible light below lasing threshold' is presented in Figure 10, where the emitted infrared light and visible light are shown with logarithmic *Y*-axis scale. Below the lasing threshold, there is spontaneous infrared emission up to 150 μW, while the visible light emission remained below the detector responsivity limit. When considering applications requiring high contrast ratio, such as near-to-eye and head-up displays, this greatly enhanced extinction ratio is expected to be of particular importance. The projected output beam of the 620-nm laser is presented in Figure 11.

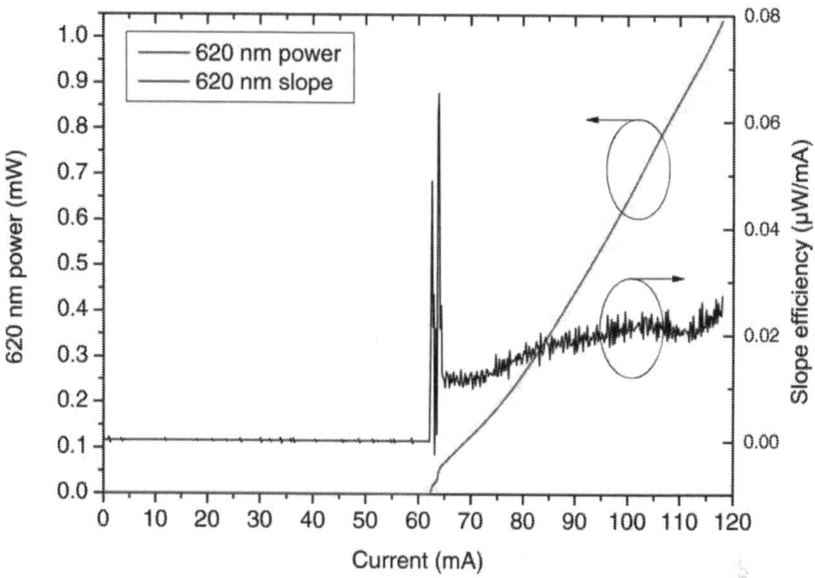

Figure 9 Frequency-converted 620-nm ***L* - *I*** performance under passively pulsed mode.

4. CONCLUSIONS

A transversally single-mode frequency-converted GaInNAs-based 620-nm laser diode is demonstrated with high single pass conversion efficiency and extinction ratio. Further improvements of threshold current and conversion efficiency are expected by optimizing the laser diode manufacturing process and optical coupling configuration.

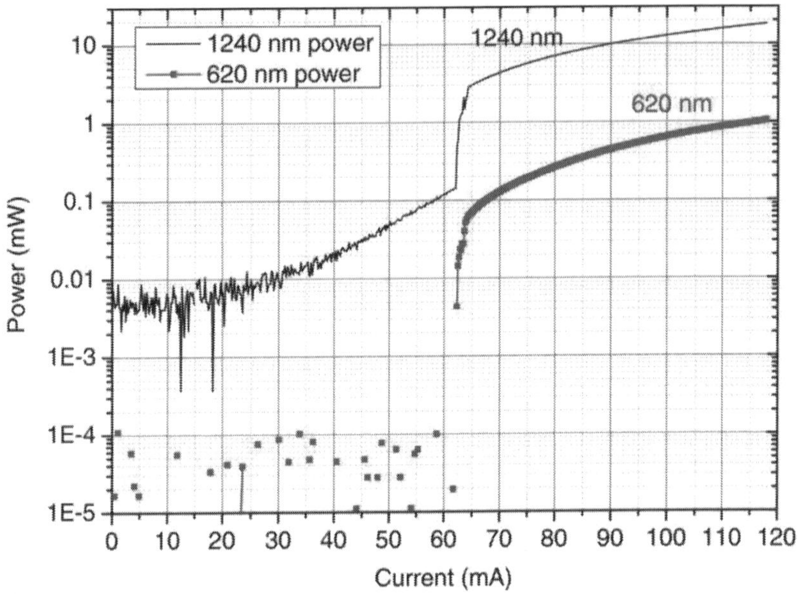

Figure 10 Comparison of frequency-converted 620-nm and infrared 1240-nm output.

Figure 11 Projected 620-nm output beam of the GaInNAs laser diode. $MgO:LiNbO_3$ nonlinear waveguide crystal was used for single-pass frequency conversion from 1240 to 620 nm.

AUTHORS' INFORMATION

JK is CTO at EpiCrystals. VMK is a PhD student at the Optoelectronics Research Centre of Tampere University of Technology.

ACKNOWLEDGEMENTS

Authors wish to thank Prof. Mircea Guina for the support in proofreading of the manuscript as well for the numerous helpful comments. VMK acknowledges the financial support of the Graduate School of Electronics, Telecommunications and Automation (GETA) and HPY Research Foundation.

COMPETING INTERESTS

The authors declared that they have no competing interests.

AUTHORS' CONTRIBUTIONS

JK carried out the laser performance characterization and writing the manuscript. VMK carried out the molecular beam epitaxy and participated in designing the semiconductor structure and writing the manuscript. Both authors read and approved the final manuscript.

REFERENCES

1. Buckley E: Detailed eye-safety analysis of laser-based scanned-beam projection systems. *J Displ Technol* 2012, 8: 166–173.

2. Bohdan R, Bercha A, Trzeciakowski W, Dybała F, Piechal B, Sanayeh MB, Reufer M, Brick P: Yellow AlGaInP/InGaP laser diodes achieved by pressure and temperature tuning. *J Appl Phys* 2008, 104: 063105.

3. Onishi T, Inoue K, Onozawa K, Takayama T, Yuri M: High-power and high-temperature operation of Mg-doped AlGaInP-based red laser diodes. *IEEE J Quant Electron* 2004, 40: 1634–1638.

4. Qiu B, McDougall S, Yanson D, Marsh J: Analysis of thermal performance of InGaP/InGaAlP quantum wells for high-power red laser diodes. *Opt Quant Electron* 2008, 40: 1149–1154.

5. Härkönen A, Rautiainen J, Guina M, Konttinen J, Tuomisto P, Orsila L, Pessa M, Okhotnikov OG: High power frequency doubled GaInNAs semiconductor disk laser emitting at 615 nm. *Opt Express* 2007, 15: 3224–3229.

6. Nakahara K, Adachi K, Kasai J, Kitatani T, Aoki M: High-performance GaInNAs-TQW edge emitting lasers. In *20th International Semiconductor Laser Conference: September 17–21 2006; Kohala Coast, HI, USA*. Edited by: IEEE. Piscataway: IEEE; 2006:161–162.

7. Bisping D, Pucicki D, Hofling S, Habermann S, Ewert D, Fischer M, Koeth J, Forchel A: High-temperature high-power operation of GaInNAs laser diodes in the 1220–1240-nm wavelength range. *IEEE Photon Technol Lett* 2008, 20: 1766–1768.

8. Tansu N, Mawst LJ: Current injection efficiency of InGaAsN quantum-well lasers. *J Appl Phys* 2005, 97: 054502.

9. Oclaro Data Sheet HL63163DG AlGaInP Laser Diode, HL63163DG Rev1 http://www.oclaro.com/datasheets/OCDE_HL63163DG_Rev_1.pdf

10. Lin C-C, Liu K-S, Wu M-C, Ko S-C, Wang W-H: Facet-coating effects on the 1.3-μm strained multiple-quantum-well AlGaInAs/InP laser diodes. *Jpn J Appl Phys* 1998, 37: 6399–6402.

11. Pliska T, Arlt S, Matuschek N, Schmidt B, Mohrdiek S, Harder C: High power wavelength stabilized 980 nm pump laser modules operating over a temperature range of 135 K. In *14th Annual Meeting of the IEEE Lasers and Electro-Optics Society (LEOS): November 12–13 2001; San Diego, CA, USA. Volume 1*. Edited by: IEEE. Piscataway: IEEE; 2001:139–140.

12. Nishida T, Shimada N, Ogawa T, Miyashita M, Yagi T: Short wavelength limitation in high power AlGaInP laser diodes. In *Proceedings of SPIE: High-Power Diode Laser Technology and Applications IX. Volume 7918*. Edited by: Zediker MS. Bellingham: SPIE; 2011:791811–791811. -7 -7

13. Blume G, Nedow O, Feise D, Pohl J, Paschke K: Monolithic 626 nm single-mode AlGaInP DBR diode laser. *Opt Express* 2013, 21: 21677–21684.

6

A NOVEL PULSE MEASUREMENT SYSTEM BY USING LASER TRIANGULATION AND A CMOS IMAGE SENSOR

Jih-Huah Wu [1], Rong-Seng Chang [2] and Joe-Air Jiang [3,*]

[1] Department of Biomedical Engineering, Ming Chuan University, No. 5, Deming Rd., Gweishan Township, Taoyuan 333, Taiwan.

[2] Department of Optics and Photonics, National Central University, No. 300, Jung-Da Rd., Chung-Li City, Taoyuan 320, Taiwan.

[3] Department of Bio-Industrial Mechatronics Engineering, National Taiwan University, No. 1, Sec. 4, Roosevelt Rd., Taipei 106, Taiwan.

ABSTRACT

This paper presents a novel, non-invasive, non-contact system to measure pulse waveforms of artery via applying laser triangulation method to detect skin surface vibration. The proposed arterial pulsation measurement (APM) system chiefly consists of a laser diode and a low cost complementary metal-oxide semiconductor (CMOS) image sensor. Laser triangulation and centroid method are combined with the Fast Fourier Transform (FFT) in this study. The shape and frequency of the arterial pulsation can be detected rapidly by using our APM system. The relative variation of the pulse at different measurement points near wrist joint is used as a prognostic guide in traditional Chinese medicine (TCM). An extensive series of experiments was conducted to evaluate the performance of the designed APM system. From experimental results, the pulse amplitude and frequency at the Chun point (related to the small intestine) of left hand showed an obvious increase after having food. In these cases, the peak to peak amplitudes and the frequencies of arterial pulsations range from 38 to 48 μm and from 1.27 to 1.35 Hz, respectively. The height of arterial pulsations on the area near wrist joint can be estimated with a resolution of better than 4 μm. This research demonstrates that applying a CMOS image sensor in designing a non-contact, portable, easy-to-use, low cost pulse measurement system is feasible. Also, the designed APM system is well suited for evaluating and pre-diagnosing the health of a human being in TCM clinical practice.

KEYWORDS

Arterial pulsation; Centroid method; CMOS image sensor; Laser triangulation; Pulse frequency

1. INTRODUCTION

The waveform of arterial pulsation is considered a fundamental indicator for the diagnosis of cardiovascular disease, which can guide therapeutic decisions in complex clinical situations [1]. Abnormalities of the waveform shape and frequency of the arterial palpitation are indicators of certain cardiovascular disorders. Thus, how to distinguish arterial pulse waveforms without distortion has become an important issue in biomedical signal processing. In addition, pulse diagnosis is one of four kinds of diagnostic methods used in TCM clinical practice to determine the physiological condition of the patients [2]. The most commonly used clinical methods to measure the behavior of the pulse include the stethoscope, sphygmomanometer, and Doppler-based instrumentation. Recently, some practitioners in TCM use a pulse diagnosis machine or other device to record changes in the pulse [3]. A set of three pressure transducers for sensing the pulses at three locations has already been developed [4]. Lu *et al.* analyzed the harmonics of the frequency spectrum of arterial pulse waves and correlated some illness conditions to certain harmonics [5]. Hong *et al.* described a non-touch pulse measurement method based on optical interferometer [6], which could estimate skin vibration velocity. However, the devices used in the above-mentioned studies could either interfere with the measurement results because of making contact with the body, or were more costly because of having additional sensors for probing multiple points. An optical non-contacting technology, which is based on optical triangulation, is proposed in this study. Laser triangulation is a well-known method in thickness and contour measurement, and has been applied to many industrial fields [7-9]. It was also used to examine the vertical movements of the vocal folds during phonation [10].

Laser triangulation is normally used in conjunction with light centroid measuring devices, e.g. position sensing detectors (PSDs) and charge coupled devices (CCDs). Since the manufacturing technology behind the CMOS image sensors has now been advanced sufficiently to achieve good stability and low cost, the CMOS image sensors have become increasingly significant in industrial optical sensors [11-12]. The CMOS image sensors possess several programmable features including electronic exposure (ET) control, continuous or single frame capture, and progressive or interlaced scanning modes. The first of these features is very important to our experiments, especially for reducing noise and locating measurement point of laser spot.

In general, the relationship between pulse waves and physiological or healthy conditions of the tested subjects is quite complex. Such relationship might exhibit nonlinear characteristic and might also vary person to person due to characteristics of the artery, deep or shallow, healthy or hardened, etc. Basically, the more pulse waveforms are obtained, the more information including pulse rates and harmonics related to the diagnosis in TCM can be achieved. In TCM clinical practice, an experienced TCM physician can do the pulse diagnosis by palpation treatment conducted on multiple measurement points (i.e., Chun, Guan, and Chy points) to determine the relationships between the organs health and the wave patterns of pulses. These facts have been demonstrated in many literatures [13-14]. Of course, a more complex and accurate arterial pulse measurement system is necessary if more factors are taking into consideration in TCM clinical practice.

This paper describes the design of the arterial pulsation measurement (APM) system, and presents the results of tests conducted to verify the pulse measurement accuracy. The pulsation rate was derived from the frequency spectrum of the laser spot vibration, and showed great consistence with data taken from loudspeaker movement driven by a function generator at a specific frequency. Frequency validation was also conducted by comparing the experimental results with data obtained from a standard blood pressure monitor. The amplitude and frequency variation at each point measured on the tested subject's wrist is an important symptom for some illness during the patient's medical examinations in TCM clinical practice. The evaluation of pulse variation gives us some valuable information concerning about the tested subjects' health.

2. PRINCIPLES OF MEASUREMENT

The proposed APM system combines the Fast Fourier Transform (FFT), the centroid method, and the optical triangulation method. The frequency spectrum of the arterial pulse waveforms measured at the specified point is obtained by FFT method. The calculations were conducted by MATLAB 7.0 and Origin 6.0. It is possible by using the FFT theory [15] to build a variety of non-sinusoidal waveforms consisting of many sinusoidal waveforms. In other words, a non-sinusoidal waveform can be decomposed into many sinusoidal waveforms with different frequencies, amplitudes, and phases. Due to the speed limitation of the CMOS image sensor, we discuss only the fundamental sinusoidal waveform in this paper.

The laser triangulation method is simple in structure. It makes possible to measure the subject's arterial pulse waveforms in a non-contact way. The experimental data show that changes in the arterial pulse waveforms can be detected by analyzing the centroid movements of a laser spot. The changes of the centroid of the laser spot, which is measured at certain points on the wrist, can be transformed into the changes in magnitude of relative height caused by skin vibration.

The basic operation principle of the proposed APM system is described as follow. A laser diode, a laser driver, and a CMOS image sensor are used to establish an optical non-contact pulse measurement device. The laser diode emits laser light onto the measurement site of skin surface where its arterial pulsation needs to be determined. The laser spot is formed on the skin surface of the wrist of tested subject and the variation of the spot image is captured by the CMOS image sensor and then projected onto the scattered points that represent arterial pulsations. These scattered light points are processed by FFT method to determine the amplitude and frequency of arterial pulses of tested subjects. In this work, the proposed APM system adopts a simple structure based on optical triangulation. The geometrical layout of the designed APM system is depicted in Fig. 1. In Fig. 1, X represents the distance between the target and the collimated lens of laser diode and δX is the small fluctuation (i.e., the distance between measured points A and B) of skin surface due to arterial pulsation. The target distance X is measured continuously. Using a simple triangulation principle, the measured X target coordinates are mapped onto the detection position d on the CMOS sensor, as shown in Fig. 1. The target distance X is given by

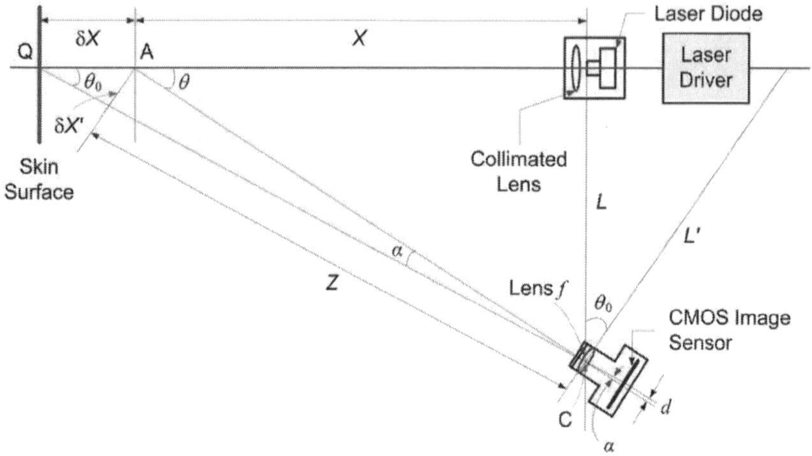

Figure 1. Geometrical layout of the arterial pulse measurement system.

$$X = \frac{L}{\tan(\theta_o + \tan^{-1}(d/f))}, \tag{1}$$

where L is the distance between the laser and the CMOS image sensor, d is the distance between the two spots mapped onto the CMOS image sensor, f is the focal length of the lens, Z is the distance between the measured point A and the center C of the lens of the CMOS image sensor, α is the angle between the axis and the measured point A, $\delta X'$ is the distance between the measured point A and the optical axis of the lens, and θ_0 is the angle between the two axes of the CMOS image sensor and the laser. In our APM system, these parameters are $X = 94$ mm, $L = 110$ mm, $Z = 144.7$ mm, and $f = 16$ mm (the focal length of the lens of the CMOS image sensor). The diameter of the lens in front of the CMOS image sensor is 6 mm.

Differentiating Eq. (1) with respect to the measured distance and rearranging the result yields

$$\delta X \cong \frac{Z^2 d}{fL}. \tag{2}$$

where δX is also regarded as the resolution of the designed APM system.

For the experiments conducted in this study, the smallest resolvable amplitude change to a sub-pixel size of d = 0.8 μm on the CMOS image sensor can be achieved. After calibration, this value corresponds to a measurement resolution of 9.5 μm achieved by the designed APM system, i.e., δX = 9.5 μm in Eq. (2). Such a measurement resolution is sufficient to detect the vibration of human pulsation.

The actual implementation of the APM system is shown in Fig. 2. The sampling area in Fig.2 was located by a TCM physician. Using simple triangulation method, the displacement δX of the variation of the laser spot mapped onto the CMOS image sensor can be determined. The amplitude and frequency of the arterial pulse can be obtained by analyzing the spot position.

The centroid method has been widely used to locate a light spot with respect to various types of image features, to sub-pixel accuracy [16-17]. The resolution of the APM system can be increased by introducing sub-pixel processing technique [16-17]. The threshold level of image pixel is deliberately set so that in the captured laser spot image a pixel with intensity below the threshold level will be neglected. In our experiments, the threshold of the gray level of image pixel is set to 20% below the highest full range level. Also, note that the vibration frequency of arterial pulsation can be obtained by processing the recorded image by means of FFT method.

A CMOS image sensor with a 5.3 × 3.8 mm^2 active area (HV7131D, manufactured by Hynix Semiconductor Incorporated.) was used to detect the laser light spot with high accuracy and stability. The sensor has a 648 × 488 pixels array and each compact active pixel element has high photo-sensitivity (3150 mV/lux-sec). It can convert the photon energy to analog voltage signal with a resolution of 8 μm. The CMOS sensor utilizes three On-chip 8-bit Digital to Analog Converts (DAC) and 648 comparators to digitize the pixel output.

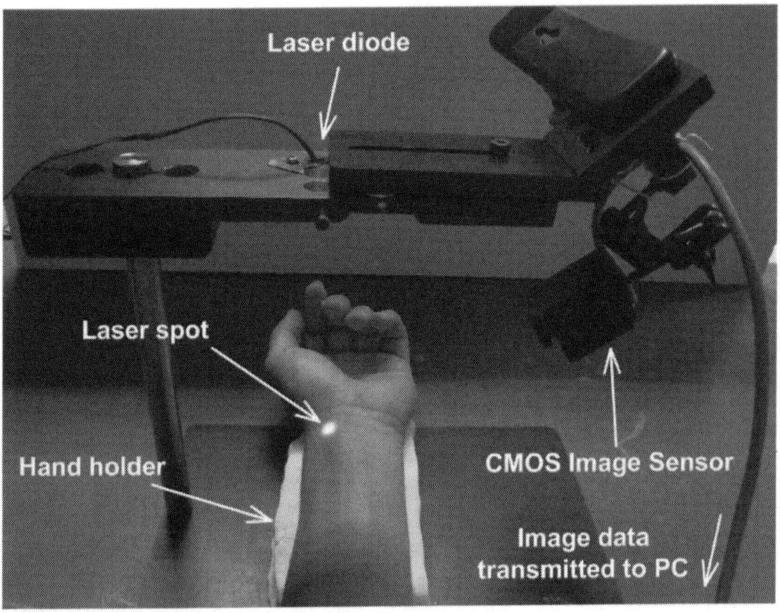

Figure 2. The actual implementation of the proposed APM system.

The output power of the adopted laser diode (Model no.: QL63d5sA, MORETEC, Inc.) is 1.3 mW, its wavelength is 650 nm, and its spectral width is about 20 nm. The diameter of the laser spot on the skin surface is approximately 1 mm, giving a spot on the CMOS sensor of tens of pixels in both directions.

At normal incidence of laser light, about a 4 ~ 7% power reflection occurs due to the differences in the refractive indices of the skin layers [18]. Also, some photons are scattered by superficial skin. The light scattered from the skin surface is the most important signal for this pulse measurement. However, photons penetrating into skin layers are also scattered, providing diffuse reflections which are of lower power than the first kind of light scattered by the superficial skin. The exposure time (ET, equivalent to gain level) of the CMOS sensor can be adjusted to different levels for different situations. This means that the signal-to-noise ratio (SNR) can be improved by setting the gain level.

There is a large amount of stray light due to the diffusely scattering nature of the skin tissue. By decreasing the ET of the CMOS sensor, the undesirable light can be eliminated. For example, Fig. 3(a) shows the experimental result when the ET value of the CMOS image sensor was set to 40. In the case of Fig. 3(b), the ET value of the CMOS image sensor was set to 1. The contours of the laser spot shown in Fig. 3(b) are obviously smaller than that in Fig. 3(a). In a normal image capturing situation, as shown in Fig. 3(a), the previously measured point can be approximately relocated by comparing the two consecutively captured frames. The signal processing flowchart is shown in Fig. 4. The software packages of MATLAB 7.0 and Origin 6.0 were used to develop the signal processing program. The program we developed can calculate the amplitude and frequency of the arterial pulse of the tested subject via examining the relative movements of the measured laser spots.

3. CALIBRATION AND VALIDATION OF THE ARTERIAL PULSE MEASUREMENT SYSTEM

The image data were recorded by a CMOS image sensor and transmitted to a personal computer for further analysis. The images were saved in bitmap format for later image processing and frequency spectrum analysis. Fifteen frames per second were captured by the CMOS image sensor. The pictures captured by the CMOS image sensor lasted for 10 or 20 seconds in each measurement. The more frames are recorded; the better resolution of the image data can be achieved.

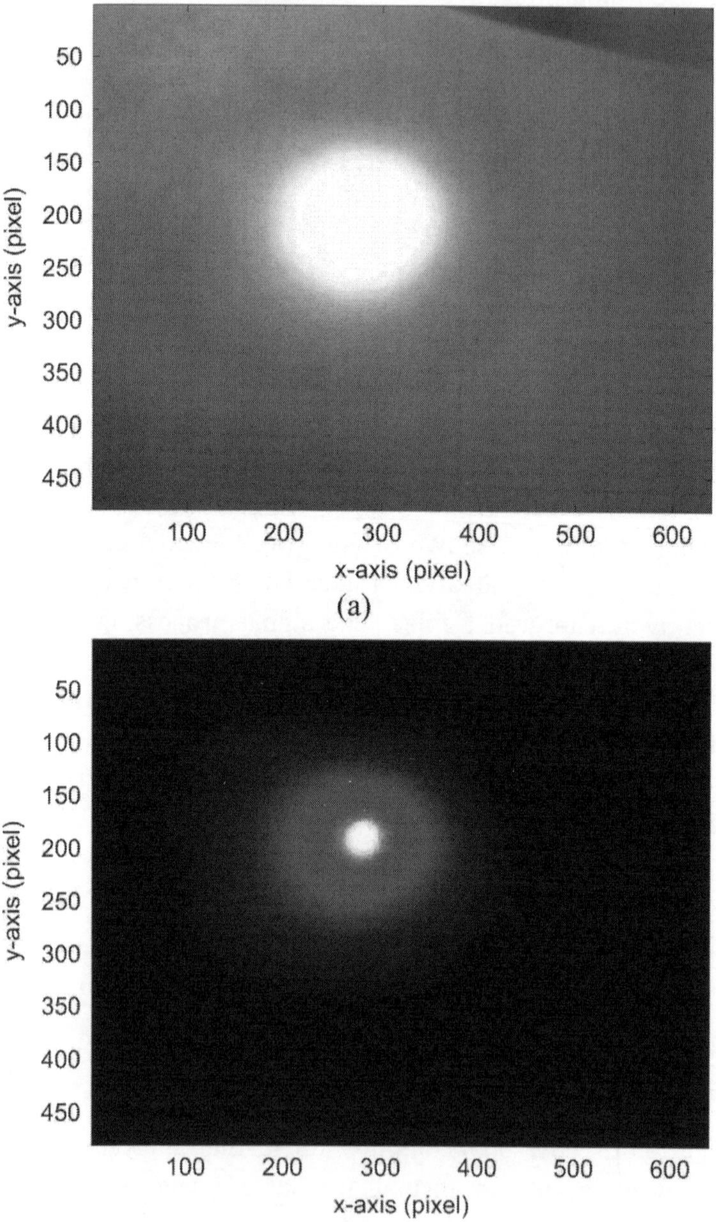

(b)

Figure 3. The contours of laser spots after ET values of the CMOS image sensor were set to: (a) 40 and (b) 1.

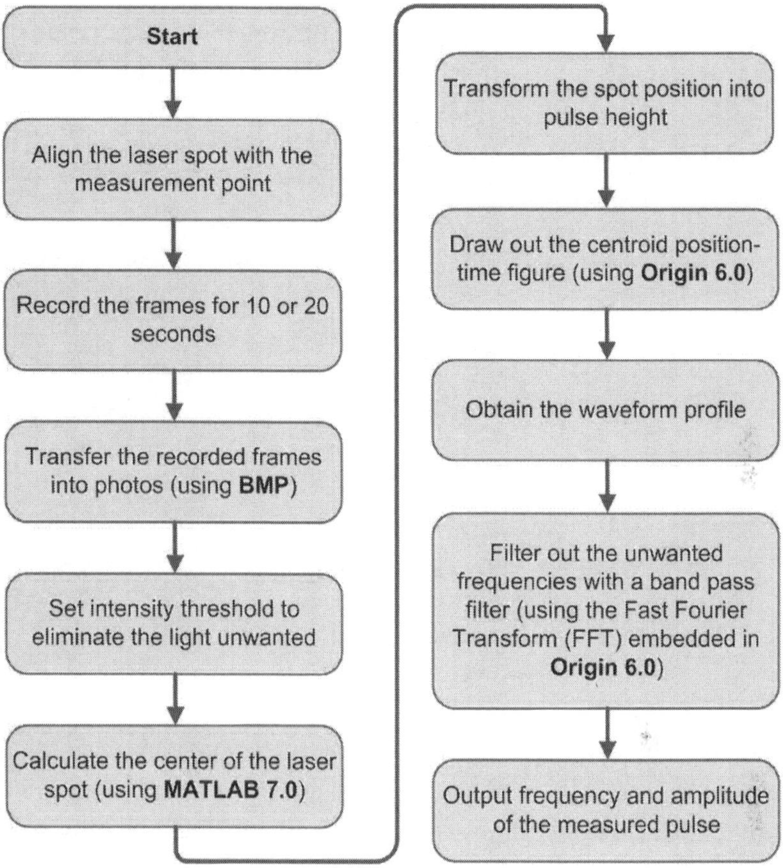

Figure 4. Signal processing flowchart for the proposed arterial pulse measurement system.

The calibration of the APM system was very straightforward and easy. A precise translator was used to calibrate the linearity of the APM system. A white paper was placed above the translator as a reference panel, and then the shift of the laser spot can be calibrated by adjusting the elevation of the translator step-by-step. The adopted step size of the translator that can mimic pulsation amplitude measurement is 20 μm. The schematic drawing of the linearity calibration of the APM system is shown in Fig. 5. The amplitude calibration of pulsation height measured by the APM system was conducted on an isolated optical table. The results for the linearity calibration experiments are shown in Fig. 6(a).

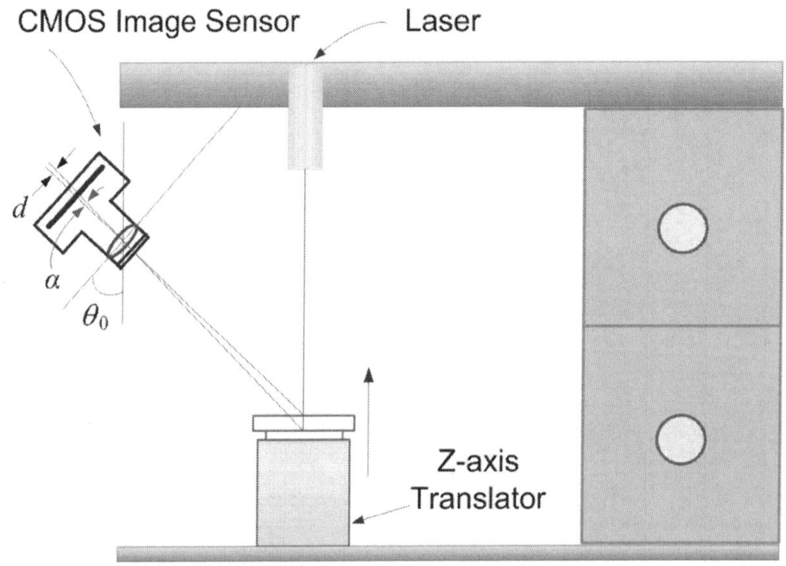

Figure 5. The schematic drawing of the linearity calibration of the APM system.

The one standard deviation (1σ) is obtained by conducting the measurements 30 times, and the results are as shown inFig. 6(b). Examining Fig. 6(b) indicates that the one standard deviation of the APM system achieved is less than 0.04 pixels, i.e., approximately equivalent to 3.8 μm. This fact demonstrates that the APM system provides pretty good performance on measurement stability.

The frequency calibration was conducted by comparing the experimental results with the drumhead movements of a loudspeaker driven by a high-precision function generator (LFG-1300, Leader, Inc.). In this experiment, we used a function generator with a specific frequency to drive the loudspeaker. So, the drumhead variation of the loudspeaker referred to a standard frequency can be provided as the reference of the frequency calibration of our APM system. The schematic drawing of the frequency calibration of the APM system is shown in Fig. 7. For the loudspeaker operated at 1.0 Hz, the amplitude variation of the loudspeaker drumhead movements in the time domain and its frequency spectrum are shown inFig. 8(a) and 8(b), respectively. The calibration results, from 0.6 to 2.0 Hz, are shown in Table 1. In this test, the frame capture rate of CMOS image sensor of the APM system is set to 15 frames/sec. The accuracy of pulse measurement could reach 2.5%, i.e. the error in the pulse rate would be less than 1.5 pulses per minute.

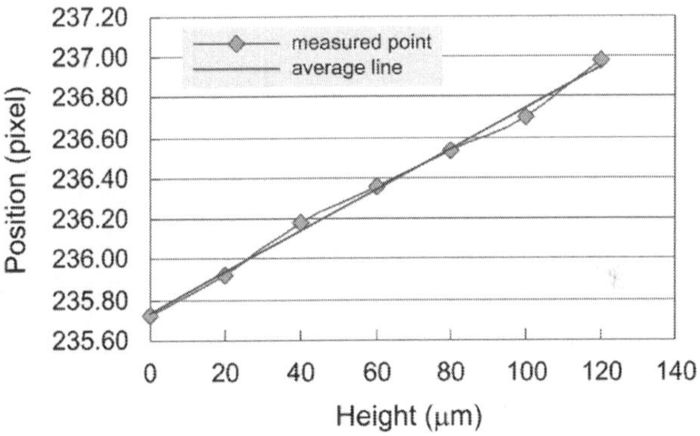

(a)

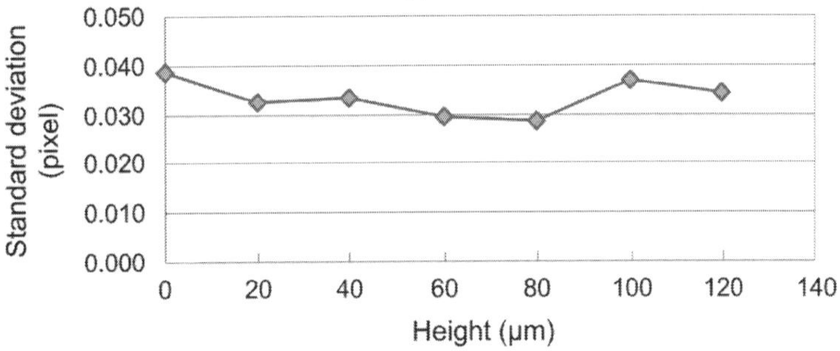

(b)

Figure 6. Calibration for the pulsation amplitude measurement of the proposed APM system: (a) the results for linearity calibration experiments and (b) the one standard deviation of amplitude calibration experiments.

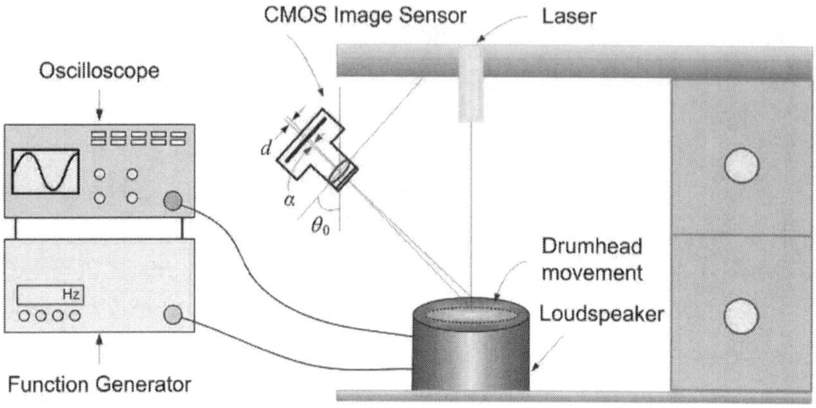

Figure 7. The schematic drawing of the frequency calibration of the APM system.

The accuracy of the frequency measurement of our APM system was also validated by comparing the experimental results with the data obtained from a blood pressure monitor (Model No.: OS-512, OSIM, Inc.). Ten healthy volunteers participated in this study. The experimental results are summarized in Table 2. The difference between the measured results obtained from the APM system and the blood pressure monitor was no more than 2.8%, i.e., 2 pulses/min.

The position variations of the laser spot centers measured in the time domain are illustrated by the thin line in Fig. 9(a). This variation curve contains a full region of information associated with vibration frequencies and amplitudes of the arterial pulses, breathings, hand movements, and involuntary body tremors. Generally, the human pulse is about 0.7 to 2 Hz. The measurement data after being filtered with a band pass filter is depicted by a thick line as shown in Fig.9 (a). The amplitude of vibration of laser spot center (in terms of pixels) has been enlarged in Fig. 9(b). It can be seen that the change of the center of the laser spot is proportional to the change in distance to the skin as calculated in Eq. (1). In Fig. 9(b), the peak to peak values of the spot center variation curve are almost fallen within 0.6 pixels, i.e., the arterial pulse amplitude is approximately equal to 57 μm. The full spectrum in frequency domain for Fig. 9(a) is shown in Fig. 10 (a). An inspection of both Fig. 9(a) and Fig. 10(a) indicates that there was low frequency vibration caused by noise, hand movements, and breathings. After filtering, the pulse frequency (1.17 Hz) could be obtained easily, as seen in Fig. 10(b).

4. EXPERIMENTAL RESULTS

In TCM there are four diagnostic methods: inspection, auscultation, questioning, and palpation. Among these methods, the pulse diagnosis by palpation is the most important and also the most difficult one. According to Chinese medical literature [13], there are in total of 29 wave patterns of the arterial pulse, each having a specific name. However, it is too difficult and too subjective for most people to distinguishing 29 different wave patterns with the finger tips. The TCM physician usually needs to use an auxiliary device to determine these patterns. The relationships between the organs health and the measurement points are illustrated in Fig. 11. This is especially meaningful in TCM clinical practice, because for pulse diagnosis the physician places his index, middle, and ring fingers on the patient's wrist, in accordance with the three locations called Chun, Guan, and Chy [14]. In our experiments, we found that variations in the pulse at relative measurement points on the wrist mentioned-above could reveal something about the physiology of the tested subjects.

Some of the measurement results for pulse amplitude and frequency obtained under different conditions are shown inFigs. 12 and 13, respectively. In Fig. 12, the pulse amplitude and frequency measured at the Chun point of left hand (related to the small intestine) showed an obvious increase after having food. Fig.12 (a) shows results observed before the meal andFig. 12(b) are those after meal, respectively. The measurements were conducted 30 minutes before and 30 minutes after eating the meal.

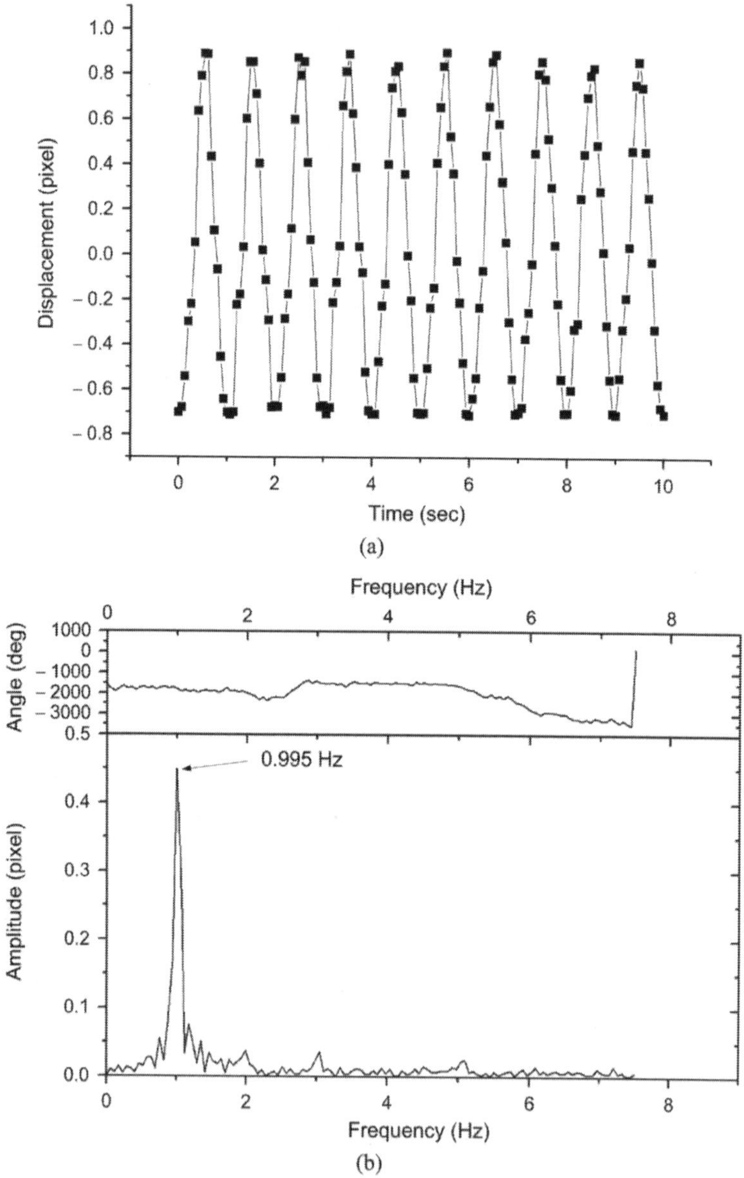

Figure 8. Frequency calibration of the APM system: (a) the spot variation versus time (time domain) and (b) the frequency spectrum analysis of the data in (a) using FFT method.

Table 1. Frequency calibration of the APM system from 0.6 – 2.0 Hz with a function generator and a loudspeaker.

Standard frequency (Hz)	Frequency obtained from the APM system	Error (%)
0.6	0.585	2.5
0.7	0.703	0.4
0.8	0.820	2.5
0.9	0.879	2.3
1.0	0.995	0.4
1.1	1.113	1.2
1.2	1.230	2.5
1.3	1.288	0.9
1.4	1.406	0.4
1.5	1.523	1.5
1.6	1.581	1.2
1.7	1.698	0.2
1.8	1.816	0.9
1.9	1.874	1.4
2.0	1.991	0.4

Table 2. Comparison of the frequency measurements made by the APM system and a blood pressure monitor (Model No.: OS-512, OSIM, Inc.).

Volunteer No.	OSIM OS-512 (pulse rate)	APM system (pulse rate)	Error (%)
1	69	68	−1.4
2	76	77	1.3
3	75	76	1.3
4	84	86	2.4
5	90	92	2.2
6	82	81	−1.2
7	78	78	0.0
8	72	74	2.8
9	94	96	2.1
10	78	78	0.0

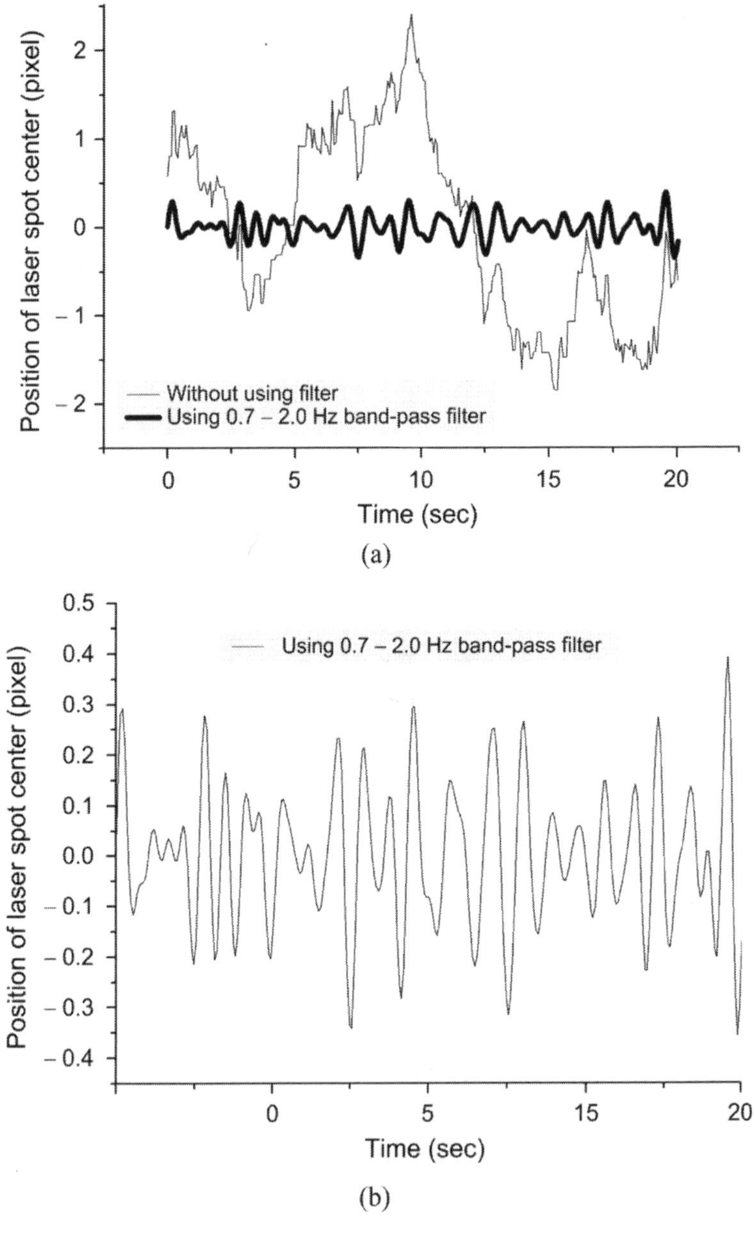

Figure 9. Laser spot centers measured by APM system: (a) the original data of centroid variation of laser spot center and (b) the data after enlarged amplitude scale and with filtering.

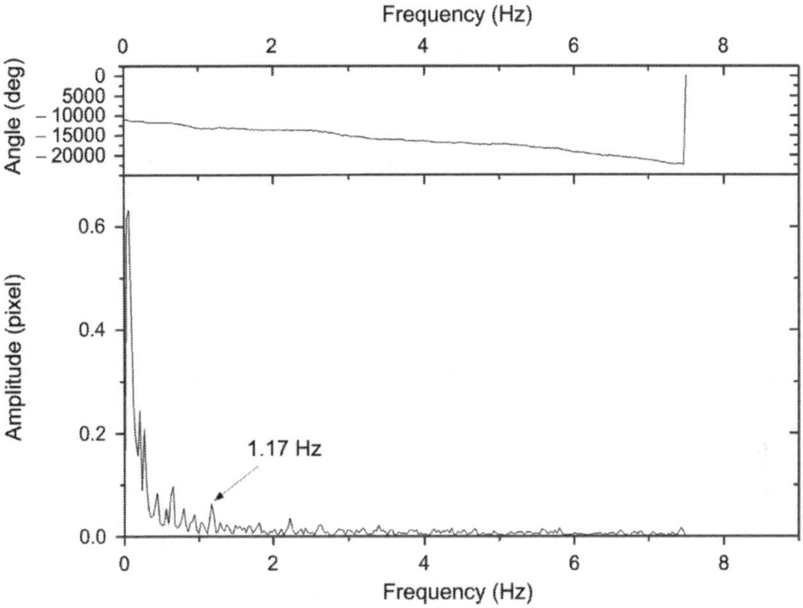

(a)

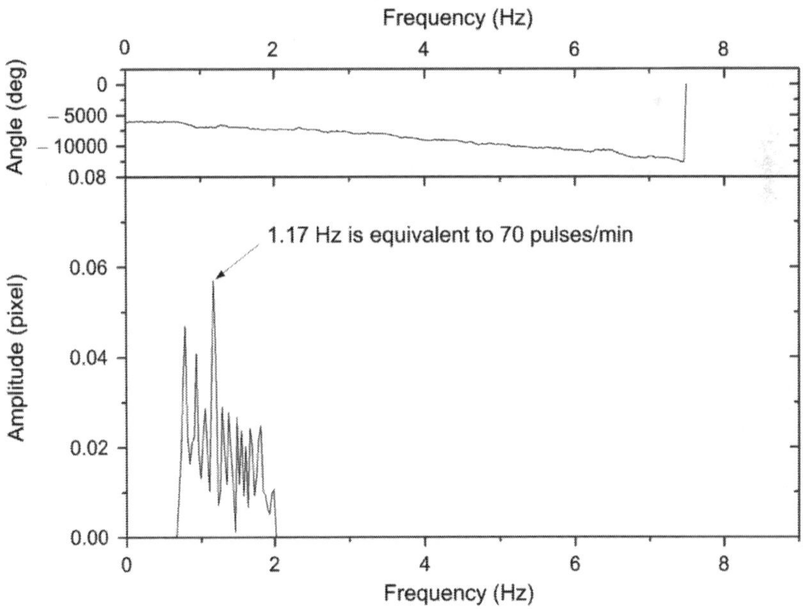

(b)

Figure 10. Full spectrum analysis of the data in Figure 9: (a) the results of without using filter and (b) the results of using filter.

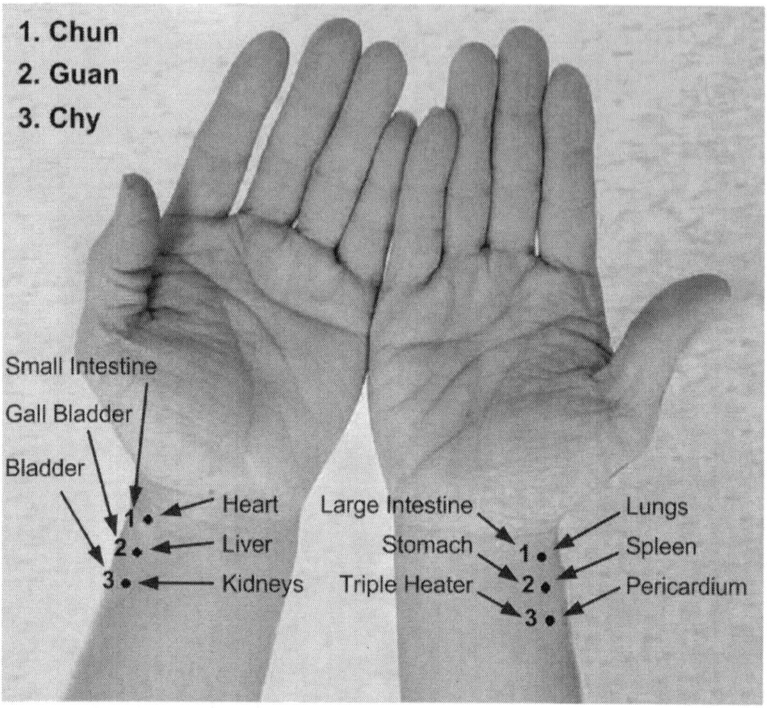

Figure 11. Illustration of palpation positions for pulse diagnosis used in traditional Chinese medicine.

In Fig. 12(a), the peak to peak value of pulsation amplitude is approximately within 0.4 pixels, i.e., the maximum variation in pulse amplitude is approximately equal to 38 μm. In Fig. 12(b), the peak to peak value of pulsation amplitude is approximately ranged from –0.32 to 0.28 pixels. This means that the maximum variation in pulse amplitude of the tested subject after meal 30 min will increase to 48 μm. An examination of Figs. 12(a) and 12(b) also observes that the pulse frequency measured at the Chun point of left hand of the tested subject is changed from 1.27 to 1.35 Hz.

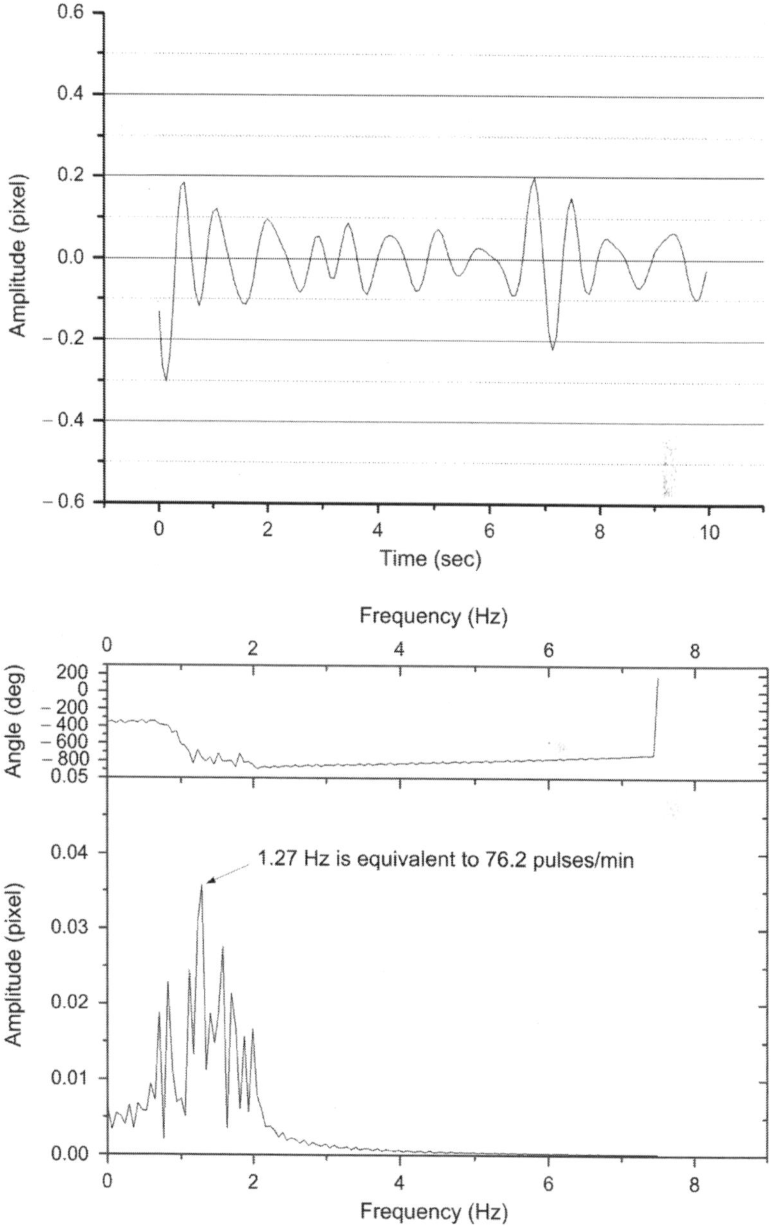
0.6
0.4
0.2
0.0
-0.2
-0.4
-0.6
Amplitude (pixel)
0
2
4
6
8
10
Time (sec)
Frequency (Hz)
Angle (deg)
200
0
-200
-400
-600
-800
0.05
0.04
0.03
0.02
0.01
0.00
Amplitude (pixel)
1.27 Hz is equivalent to 76.2 pulses/min
Frequency (Hz)

(a)

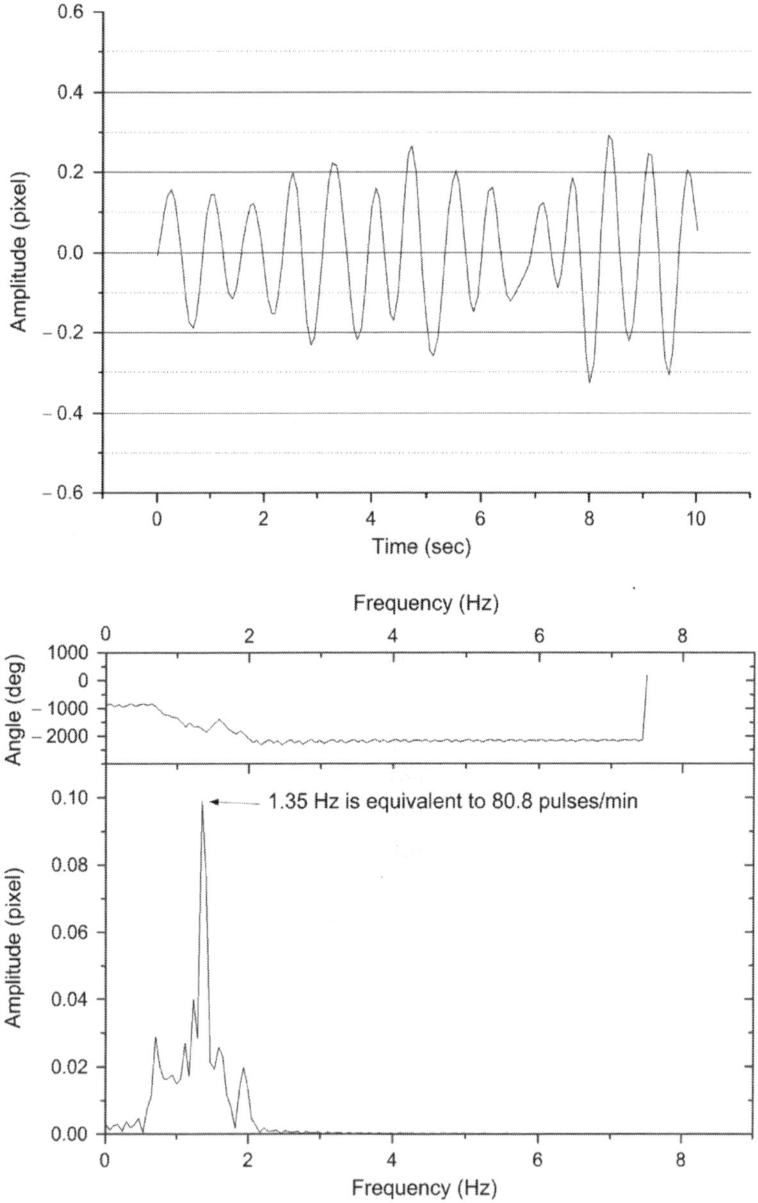

(b)

Figure 12. Pulse information measured at the Chun point on the left hand (small intestine) of the tested subject: (a) 30 min before a meal, (b) 30 min after a meal.

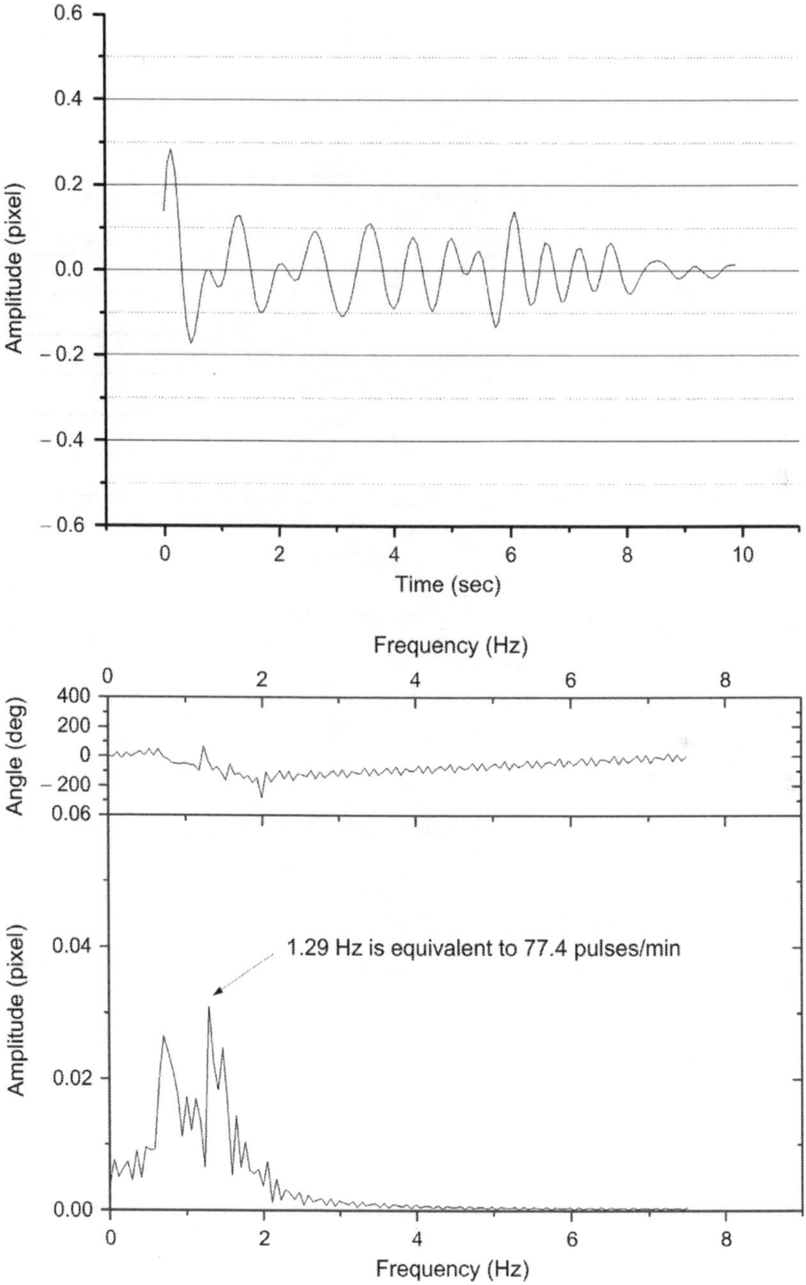

0.6
0.4
0.2
0.0
−0.2
−0.4
−0.6
Amplitude (pixel)
0
2
4
6
8
10
Time (sec)
Frequency (Hz)
0
2
4
6
8
400
200
0
−200
0.06
Angle (deg)
0.04
0.02
0.00
Amplitude (pixel)
1.29 Hz is equivalent to 77.4 pulses/min
0
2
4
6
8
Frequency (Hz)

(a)

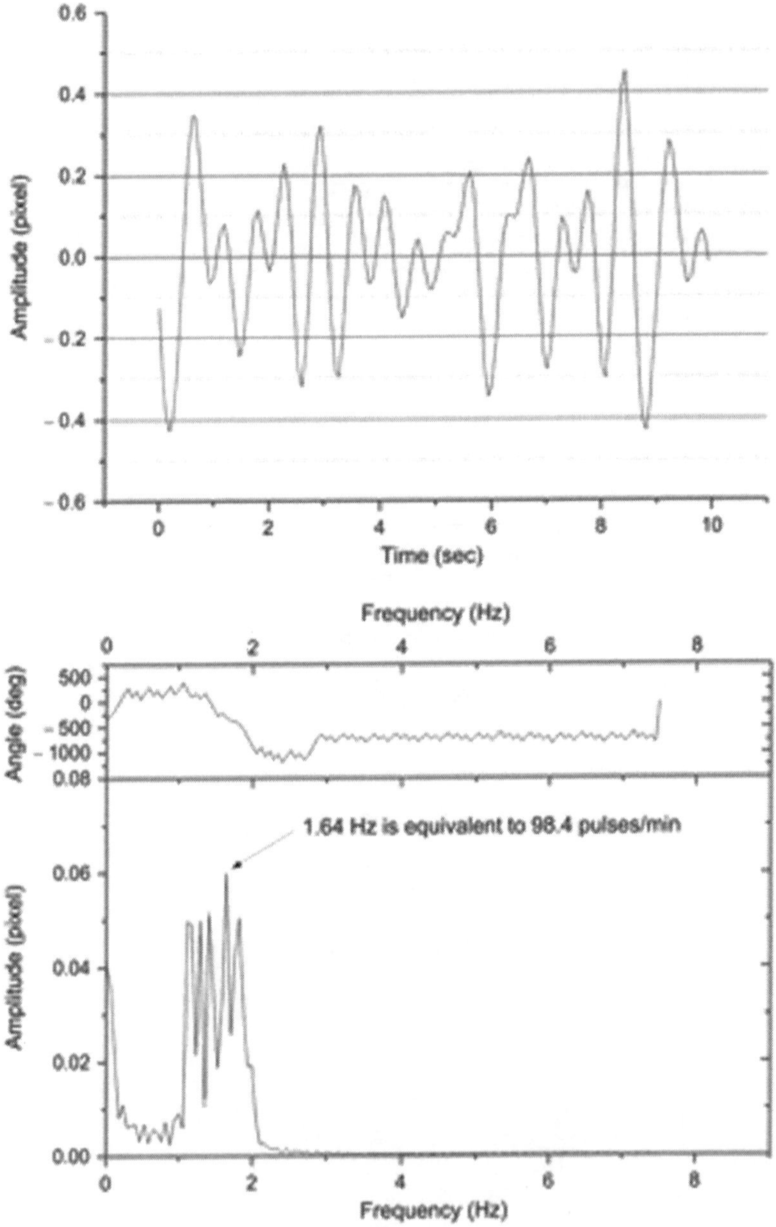

(b)

Figure 13. Pulse information for the Guan point on the left hand (liver) of the tested subject: (a) before staying up late, (b) after staying up late.

The other test showed that staying up late caused changes in the amplitude and frequency of the pulse at the Guan point of left hand. The pulsation at the Guan point of left hand is closely related to the liver activity. Without staying up late, the amplitude and frequency of the pulsation of the tested subject measured at Guan point were normal, as shown in Fig. 13(a). But after staying up late, it showed an apparent increase in the amplitude of measurement data, as shown in Fig. 13(b). InFig. 13(a), the peak to peak amplitude of pulsation measured at the Guan point on the left hand of the tested subject varies in small range, approximately equal to 0.2 pixels (i.e., equivalent to 19 μm). In Fig. 13(b), for the tested subject staying up late the pulse variation in peak to peak amplitude measured at the Guan point enlarged approximately to 0.6 pixels (for most portion), which is equivalent to 57 μm. In this case, the pulse frequency is also changed from 1.29 to 1.64 Hz. In TCM practice, this measurement indicated the state of the liver's health.

5. CONCLUSIONS

Based on optical laser triangulation theory, a non-invasive and non-contact arterial pulsation measurement (APM) system to detect micro-vibration on skin surface is developed in this work. The APM system consists chiefly of a laser diode and a CMOS image sensor, and the implementation cost is pretty low. An extensive series of experiments to evaluate the performance of the APM system was conducted. The pulse waveforms of the tested subject can be detected by our APM system easily. The APM system achieves a measurement resolution of μm order. Experimental results also show that the amplitude and frequency of the pulse of tester have been changed under different conditions. These tests demonstrate the performance of the proposed APM system for measuring micro-pulsation on skin surface is pretty good. If a speedier CMOS or CCD image sensor, such as 200 frames per second or more and a smaller pixel size can be used, the pulse waveform obtained by our APM system would be more accurate and clearer. To reduce the speckle effect of the laser, a non-coherent light could be chosen. This would also reduce the noise caused by laser light.

Our experimental results have shown the feasibility of using the optical triangulation method and a CMOS image sensor to measure arterial pulsation. Although the demonstrated examples are not yet sufficient to clinical bearings, they serve to test the method and evaluate the performance of the proposed APM system. In the future, this arterial pulse measurement system can be improved by using 3 light sources to simultaneously check the pulsation of three or six different TCM points on the wrist (i.e., Chun, Guan, and Chy points on one hand or two hands, respectively). And the relation between the pulse signal and the healthy condition of the subject will be established. We hope that we can report the investigation results in future.

ACKNOWLEDGMENTS

The authors are grateful to the National Science Council of the Republic of China for financially supporting this research under contract no.: NSC 94-2213-E-002-120 and NSC 95-2218-E-002-073.

REFERENCES AND NOTES

1. Martin, J. F.; Volfson, L. B.; Kirzon-Zolin, V.V.; Schukin, V.G. Application of pattern recognition and image classification techniques to determine continuous cardiac output from the arterial pressure waveform. *IEEE Trans. Biomed. Eng.* **1994**, *41*(10), 913–920.

2. Fu, S. E.; Lai, S. P. A system for pulse measurement and analysis of Chinese medicine., Proceedings of IEEE Engineering in

Medicine & Biology Society's 11th Annual International Conference; 1989; pp. 1695–1696.

3. Min, S. G.; Yoon, D. H. Convergence characteristics of MADF algorithm in the pulse diagnostic apparatus. Proceedings of IEEE Region 10 Conference; 1999; 2, pp. 1551–1554.

4. Yoon, S. H.; Koga, Y.; Matsumoto, I.; Ikezono, E. An objective method of pulse diagnosis. *American J. of Chinese Medicine* **1987**, *15*(3/4), 147–153.

5. Lu, W.A.; Lin, Y.Y.; Wang, W. K. Pulse analysis of patients with severe liver problems. *IEEE Eng. in Med. and Bio.* **1999**,*18*(1), 73–75.

6. Hong, H. D.; Fox, M. D. No touch pulse measurement by optical interferometry. *IEEE Trans. Biomed. Eng.* **1994**, *41*(11), 1096–1099.

7. Sequeira, V.; Goncalves, J. G. H.; Isabel Ribeiro, M. 3D environment modeling using laser range sensing. *Robotics and Autonomous Systems* **1995**, *16*, 81–91.

8. Amann, M. C.; Bosch, T.; Myllylä, R.; Rioux, M. Laser ranging: a critical review of usual techniques for distance measurement. *Opt. Eng.* **2001**, *40*(1), 10–19.

9. Wu, J. H.; Horng, K. Y.; Chang, R. S. Triangulation-based range finder using a complementary metal-oxide semiconductor (CMOS) image sensor for application to automotive systems. *Opt. Eng.* **2006**, *45*(4).

10. Manneberg, G.; Hertegard, S.; Liljencrantz, J. Measurement of human vocal fold vibrations with laser triangulation.*Opt. Eng.* **2001**, *40*(9), 2041–2044.

11. Agwani, S.; Domer, S.; Rubacha, R.; Stanley, S. High performance, VGA resolution digital color CMOS imager. In SPIE 3649,

Conference on Sensors, Cameras, and Systems for Scientific/Industrial Applications; 1999; pp. 154–164.

12. Wong, H. S. Technology and device scaling considerations for CMOS imagers. *IEEE Trans. on Electron Devices* **1996**,*43*(12), 2131–2342.

13. Wang, S. H. 265-317AD; *The classic of the Pulse (in Chinese).*; reprinted by Wen Kuan Publishing Co.: Taipei, Taiwan, ROC, 1972.

14. Dharmananda, S. *The significance of traditional pulse diagnosis in the modern practice of Chinese medicine.* Institute for Traditional Medicine: Portland, Oregon, 2000. http://www.itmonline.org/arts/pulse.htm.

15. Ramirez, R. W. *The FFT Fundamentals and Concepts*; Tektronix Inc.: Englewood Cliffs, New Jersey, 1986; pp. 12–13.

16. Dorsch, R. G.; Häusler, G.; Herrmann, J. M. Laser triangulation: fundamental uncertainty in distance measurement.*Appl. Opt.* **1994**, *33*(7), 1306–1314.

17. Fillard, J. P. Subpixel accuracy location estimation from digital signals. *Opt. Eng.* **1992**, *31*(11), 2465–2471.

18. Van Gemert, M. J. C.; Jacques, S. L.; Sterenborg, H. J. C. M.; Star, W. M. Skin optics. *IEEE Trans. Biomed. Eng.* 1989,*36*(12), 1146–1154.

7

CONTINUOUS IN SITU MEASUREMENTS OF VOLCANIC GASES WITH A DIODE-LASER-BASED SPECTROMETER: CO_2 AND H_2O CONCENTRATION AND SOIL DEGASSING AT VULCANO (AEOLIAN ISLANDS: ITALY)

Maurizio De Rosa[1], Gianluca Gagliardi[1], Alessandra Rocco[1], Renato Somma[2], Paolo De Natale[1] and Giuseppe De Natale[2]

[1]CNR, Istituto Nazionale di Ottica Applicata, Sezione di Napoli, Via Campi Flegrei 34, I-80078 Pozzuoli (NA), Italy

[2]INGV, Osservatorio Vesuviano, Via Diocleziano 328, I-80124 Napoli, Italy

ABSTRACT

We report on a continuous-measurement campaign carried out in Vulcano (Aeolian islands, Sicily), devoted to the simultaneous monitoring of CO_2 and H_2O concentrations. The measurements were performed with an absorption spectrometer based on a semiconductor laser source emitting around a 2-μm wavelength. The emitted radiation was selectively absorbed by two molecular ro-vibrational transitions specific of the investigated species. Data for CO_2 and H_2O concentrations, and CO_2 soil diffusive flux using an accumulation chamber configuration, were collected at several interesting sampling points on the island (Porto Levante beach- PLB, Fossa Grande Crater – FOG- and Valley of Palizzi, PAL). CO_2/H_2O values, measured on the ground, are very similar (around 0.019 (± 0.006)) and comparable to the previous discrete detected values of 0.213 (Fumarole F5-La Fossa crater rim) and 0.012 (Fumarole VFS – Baia Levante beach) obtaid during the 1977–1993 heating phase of the crater fumaroles.

In this work much more homogeneous values are found in different points of the three sites investigated. The field work, although carried out in a limited time window (25th–28th August 2004), pointed out the new apparatus is suitable for continuous gas monitoring of the two species and their ratios, which are important geochemical indicators of volcanic activity, for which other reliable continuous monitoring systems are not yet available.

1. BACKGROUND

Continuous monitoring of gases released in volcanic areas, emitted as fumaroles or diffusive degassing from soils, can give an insight for a deeper understanding of the volcanic dynamics and its impact on global climate and human activities [1, 2]. The presence of numerous chemical species released in air and their variation vs. time can be related to chemical reactions of magma with host rocks and circulating fluids, occurring during its uprise. CO_2 and H_2O are molecular species which play a major role in volcanic dynamics. In fact, while water vapour can be related to an increasing heat flow caused by magma uprise movements, the volcanic CO_2 is in general very abundant in magma but is less reactive than other gases during the interaction with acquifers and rocks [3]. The selection of these two species is based on their common use in volcanology and on their potential in providing useful information regarding volcano degassing dynamics. The CO_2 is hence the best indicator of new deep magma supply, because its increase in concentration and flow at ground reflects increased exsolution from deep sources. In addition, since the CO_2 is by far the most abundant gas component in volcanic fluids, the ratio between CO_2 and H_2O closely represent the gas/water ratio in the aquifers, which is sensitive to temperature increase in the deeper systems, which can result, for instance, by rising of hot fluids or magma [4].

Even though in the past years different techniques have been applied to monitoring of volcanic gases, e.g., mass spectrometry and gas cromatography, they are generally performed in laboratory, after collecting sample gases on field. No reliable techniques for long-term, continuous geochemical monitoring with near-laboratory precision do actually exist. Discrete on-site field measurements are generally restricted to CO_2 soil diffusive flux [5]. Remote sensing techniques were developed in the last years for several gas species, such as Differential Optical Absorption Spectroscopy (DOAS) [6], Correlation Spectroscopy (COSPEC) [7], Light Detection And Ranging (LIDAR) [8], Fourier Transform Infrared (FTIR) Spectroscopy [9, 10]. Eddy-covariance [11] (EC) micrometereological method can be alternative solutions, for CO_2 flux, allowing measurements over large gas emission areas. Some of these methods are able to detect few trace species (e.g. SO_2, HCl, HF) while H_2S, CO_2 and H_2O, main constituents of volcanic gases, are spectroscopically hard to determine, therefore the corresponding volcanogenic emission rates can only be derived indirectly. More recently, continuous gas and temperature measurements have been performed using a set of analytical instruments to monitor volcanic gases over weeks [12–14].

Laser spectroscopy has been shown to be a powerful technique for field, non-invasive and real-time gas detection, based on the selective absorption of laser radiation by the molecules of the target gas [15, 16]. Most molecules of interest have fundamental and overtone absorption bands in the infrared region of the electromagnetic spectrum. However, in the spectral region below 2μm the outstanding progress in the telecommunication technologies made available laser sources, like distributed feed-back (DFB) semiconductor lasers, which are ideal for the development of high-resolution, fast-response spectrometers. In previous works we applied this technology to monitor CO_2 and H_2O in volcanic fumaroles, using a dual-wavelength approach [17], delivering a beam with 1.57 and 1.39 μm diode-laser radiation to a 20-cm-long cell placed upon the emitting fumarole by means of optical fibers. The dual-wavelength spectrometer enabled us to measure water vapor concentration with a 3% accuracy, whereas it was quite inaccurate for carbon dioxide measurements [18]. Subsequently, we implemented a new absorption spectrometer using a single room-temperature 2-μm DFB diode laser for real-time measurement of both CO_2 and H_2O concentration [19, 20]. At this wavelength, a considerably larger fractional absorption can be detected, due to the occurrence of stronger ro-vibrational transitions. The measurement technique was based on recording of direct absorption profiles of a gas sample contained in a multiple-reflection cell with a path-length of 50 m. Remote operation was still feasible using telecom-type optical fibers. Laboratory tests showed that it was possible to achieve a short-term reproducibility over several tens of minutes and an overall accuracy below 1% for measurements on ambient CO_2. Simultaneous measurements of the CO_2 and H_2O concentrations mantain similar accuracy levels [20, 21].

After the first field tests [20], the spectrometer has been upgraded and in this paper, we present the results of a measurement campaign carried out in August 2004 in Vulcano (Aeolian islands, Sicily). Vulcano, due to the long period of geochemical anomalies and seismicity [22, 23], and the presence of high CO_2 degassing [24–27], represents an ideal test site for the new spectroscopic apparatus, both for concentration and flux measurements. Further, the presence at some sites of fumaroles with very aggressive (hot and acid) components allow to test the apparatus and optics also in extreme conditions of operation. The volcanic CO_2 discharges occur both from fumaroles and as diffuse soil emanations form the volcanic edifices [28].

2. GEOLOGICAL SETTING OF VULCANO FUMAROLES AND SOIL DEGASSING COMPOSITION

The Vulcano Island is one of the most active volcanoes of the Mediterranean area; since its last explosive eruption, which occurred in 1888–1890 from the Fossa crater, the volcano has experienced fumarolic activity, mostly located along the northern rim of this crater [22]. Owing to a potentially pre-eruptive crisis begun in late 1977, numerous papers have been focused during the last 25 years on understanding the volcanic system at Volcano Island and on volcanic surveillance. Chemical composition changes of the fumarolic gases have been monitored for two crater fumaroles (FA and Fll) by Capasso et al. from the 1988 to 1991 [29]. The fumaroles studied showed an increase in temperature, ranging from 590°C to a maximum of 650°C for the FA fumarole. Moreover, the He/N_2, He/CO_2 and He/S ratios showed large variations with a general increase of helium, nitrogen, carbon dioxide and total sulphur contents and indicating a general trend like He>N_2 >CO_2 >total S. The behaviour of HCl concentration in the fumarole FA is opposite respect to those of He, N_2 CO_2 SO_2. In average the variability of the gas composition at Vulcano have been explained [23] as an indicator of different degree of mixing processes between a Deeper Magmatic Component (DMC) and a Shallow Hydrotermal Component (SHC). Taking into account both non-reactive gases (e.g., N_2, and He) and main constituents (i.e., H_2O, and CO_2), the maximum contribution of the DMC occurred in the summer of the 1988 while the maximum fraction of the SHC in fumarolic fluids was recorded through 1989. The following progressive increase in CO_2concentration through 1990 indicates an increasing DMC contribution, likely due to an energetic seismic event, which should have caused intense fracturation and permeability increase; it was followed by its decrease during the 1991, and finally followed, during the 1992, by an increase of the DMC contribution [23]. During the 1993–1995 period [30] monitoring of three types of sampled manifestations at Vulcano (Crater fumaroles, Baia di Levante and SIP, Palizzi and P4MAX) showed the occurrence of mixing between different end-members (crater hottest fumaroles, cooler shallower fluids, atmospheric component).

Gurrieri and Valenza have reported a diffusive CO_2 gas monitoring at Vulcano island inserting in the soil a pipe opened at the base and measuring the dynamic concentration of the CO_2 which is proportional to the ϕCO2, unless a conversion factor, depending on the experimental device, working conditions, physical characteristics of the soil [31]. Since the end of 1984, several campaigns have been carried out using the latter method in the area of Vulcano Porto and in the area surrounding the cone of La Fossa [32, 33], reporting values of flux as high as $8{\cdot}10^{-4}$ cm/s. Carapezza and Diliberto [34] have measured helium and CO_2 soil degassing respectively using a mass spectrometer in the lab and using an IR spectrometer, directly in field, with an accuracy for CO_2 concentration of 100 ppm. The average concentration of CO_2 measured in the 1989–1991 in fifty-one different points shows a general increasing trend with fluctuations, from 1989 to the end of 1991, with a maximum value at the end of 1990. Moreover during this campaign it was evidenced that some points, located on the southern base of the cone (Grotta Palizzi) and on the W and NW cone slope (Forgia Vecchia) show higher values than those found in Vulcano Porto area. CO_2 fluxes measured since the 1989 [33] suggest an increasing trend since the end of 1989 ($\sim 0.9{\cdot}10^{-4}$ cm/s per area unit) to the end of 1991 ($\sim 4.9{\cdot}10^{-4}$ cm/s per area unit), decreasing during the 1992–1993 period ($\sim 1.2{\cdot}10^{-4}$ cm/s per area unit). It is worth of note that the observed temporal variations are not influenced by seasonal variation of atmospheric parameters. This means that measurements of CO_2 soil degassing can constitute an useful indicator of deep dynamics in case of further evolution of the activity in the island whereas the crater fumaroles get inaccessible. Moreover, the collected data at specific points (Point 4) during the 1990–1992 years interval seem to suggest that a feeding system was active and enriched in He and CO_2, probably connected to La Fossa volcano [35], whereas elsewhere in the surveyed area He concentration was controlled by the extent of mixing between air and exhaling CO_2.

Continuous gas monitoring of CO_2 in soils [36] carried out at different areas of the Vulcano island shows that the mean values of CO_2 degassing in the most representative areas were lower than those of the previous year, in agreement with the observation carried out in the area of Vulcano Porto during the monthly surveys [33].

Accumulation chamber methodology used during a campaign in the period April-July 1996 in the upper part of the Fossa cone [25] measured a total output of 200 td^{-1} of CO_2 which corresponds to approximately 1000 Mgd^{-1} of steam. In order to obtain the $\phi CO2$ in the plains adjacent to the western and southern slopes of the Fossa volcanic edifice and in the isthmus separating Porto di Levante from Porto di Ponente, Chiodini et al. [26] measured emissions from 0.2 to 2900 gm^{-2} d^{-1} for a total amount of diffuse CO_2 output of 75.3 Mgd^{-1}. A survey on fifteen measurements of $\phi CH4$, $\phi CO2$ and temperature was conducted at Porto di Levante [24] obtaining respectively: 0.007 to 3.9 g $m^{-2}d^{-1}$, 6.4 to 12000 $gm^{-2}d^{-1}$, and 20 to 100°C. Carapezza and Granierii [37] have proposed a comparison between active Dynamic Concentration and passive Accumulation Chamber methods to measure the carbon dioxide soil flux at the La Fossa areas of Vulcano Island. This work showed that the active methods overestimate the CO_2flux, and the results are proportional to it only in high-flux zones. Moreover, the carbon dioxide soil flux values were controlled in proximity of active gas-releasing fractures and in presence of changes in barometric pressure and soil permeability.

3. EXPERIMENTAL SETUP

The Portable Diode Laser Spectrometer (PDLS) is based on a diode laser emitting infrared radiation around 2 μm, where different ro-vibrational transition of CO_2 and H_2O occur. The PDLS has been previously described and characterized in details in [19–21]. Here we recall the principle of operation and the major features, mainly discussing the changes made in order to improve both the robustness of the system as well as its compactness, as required for a field instrument.

The spectrometer is made by three parts: the laser breadboard, the detection bench and the electronics. A 30 × 40 cm^2 ultra-light breadboard (10-kg-weight) hosts a DFB semiconductor laser (Sensor Unlimited) emitting at 1997 nm. The emitted radiation is collimated and coupled to a 30-m-long single-mode optical fiber by means of two stealing mirrors and a input lens which focus the ligth on one fiber end. In this way we are able to steadly couple about 10% of the incoming power.

With respect to the previous version of the spectrometer a factor 3 in the weight of the laser breadboard has been gained. The use of the fiber allows to keep the laser and the controlling electronics in a safe place, far from the sampling point, where environmental conditions can often be rather hostile and compromise a normal operation. The radiation is guided by the optical fiber to the detection bench, where it is collimated and injected by lenses and tilting mirrors into an open-path multipass Herriott cell. This cell is made by two gold-coated spherical mirrors, 50-mm-diameter and ~400-mm curvature radius, rigidly aligned and bolt 400 mm apart in the central part of the detection bench. All the lenses were coated with anti-reflective layer for reducing the occurrence of fringes which could deteriorate the absorption profile. A rectangular aperture (30 × 30 cm^2) at the bottom of the bench allows the sampled gas flowing between the mirrors. The total optical path of the cell is 20 m long. The radiation transmitted by the cell is finally detected by a preamplified extended InGaAs detector. Also the detection bench has been modified, as a consequence of the experience of a previous field campaign (2002–2003) [20]: all the external surfaces are made of a titanium alloy, resulting in fact not only ligther compared to stainless steel, but also more resistant to acid compounds emitted in volcanic areas. As a result of experience for the previous campaigns, it was in fact observed that the mechanical components are the ones most suffering the hostile ambient, while the optics are not severely affected even after several days of operation. In order to vary the height of the breadboard with respect to the soil, the bench is equipped with extensible legs. The sides of the detection breadboard are protected to avoid external disturbances such as wind and dust. Then, a removable cover closes the upper part and allows also accumulation measurements in the detection chamber. In this way an evaluation of gas fluxes (ϕCO2soil) emitted by the soil can be performed.

The electronics and a laptop computer (LC) operate the laser, acquire the absorption signals and process them in real time. The laser temperature is actively stabilized providing a coarse tuning of its emission wavelength, then the frequency of the laser is continously scanned by sweeping its injection current with a triangular current ramp (2-Hz scan rate) around an offset constant value. The scanned frequency range is chosen in such a way to probe simultaneously the two absorption molecular transitions of CO_2 and H_2O. The laser frequency is swept over a range of ~20 GHz which includes the two absorption features of CO_2 (line*R*(34) of the ν_1 + 2ν02 + ν_3 combination band) and H_2O (line $15_{0,15} \rightarrow 14_{0,14}$ of the ν_2 + ν_3 band). The absorption spectra are acquired by means of a multifunction PCMCIA digital acquisition board (National Instruments NI DAQCard-6036E for PCMCIA), with a 16-bit analog-to-digital converter (ADC), at a maximal acquisition rate of 200 kS/s, interfaced to the LC. The temperature of the gas is also monitored in real time, with accuracy of 0.1°C, by acquiring the signal of a thermocouple via one of the ADC channels. A single absorption signal is acquired at the frequency scan of 2 Hz, so concentration values can be obtained every 0.5 s. Otherwise, it is possible to average several signal before processing the spectra and retrieve the concentration values.

The process is made in real time, even though all the spectra are stored for further off-line analysis. The acquisition program, written in Lab View® language, controls the intruments settings and analyzes the acquired spectra. The fitting procedure described below is made by running a Matlab routine called by the main program.

4. SIGNAL ANALYSIS AND PRELIMINARY TESTS

If $P_0(\nu)$ is the power emitted by the laser as a function of the frequency ν, then the power $P(\nu)$ detected after passing through the absorbing cell is given, according to the Beer-Lambert law, by

$$P(\nu) = P_0(\nu)\exp[A_{CO_2}g(\nu - \nu_{CO_2}) + A_{H_2O}g(\nu - \nu_{H_2O})], \quad (1)$$

Here $g(\nu - \nu_X)$ is the unit-area normalized lineshape function, centred at corresponding line center frequency ν_X for the transition X, which is assumed to be Lorentzian, as the gas is usually at atmospheric pressure:

$$g(\nu - \nu_X) = \frac{1}{\pi}\frac{\gamma_X}{(\nu - \nu_X)^2 + \gamma_X^2},$$

where γ_X is the Half Width at Half Maximum (HWHM). A_X is the integral area of the absorption feature which is the observable related to the gas concentration N_X, according to $A_X = -S_X(T)\,LN_X$,

where $S_X(T)$ is the line strength of the transition, which depends on the temperature of the sample through a known relation, and L is the absorption path length. The identification of the chosen spectral lines has been made according the spectroscopical database HITRAN [38], which also excluded the presence of other interfering gas species. The frequency scale has been previously determined by calibrating the time scale of the modulating ramp with a Fabry-Pérot cavity with known free spectral range [19]. Due to the long time stability of the laser source, this calibration has been shown to be stable over several months. Eq. (1) is fitted to the absorption spectrum according to the Levenberg-Marquardt method, and from the knowledge of the parameters resulting from the fit we can calculate the integral area A_X of the absorption and finally retrieving, through Eq. (2), the concentration of the two species, knowing the two experimentally determined transition line strengths, S_{CO_2} = 4.77(3) × 10^{-22}cm/mol at 301.4 K [19] and S_{H_2O} = 2.28(4) × 10^{-24} cm/mol at 298 K [21], and the optical path length L = 20.3 m. The baseline has been assumed to be a second degree polynomial, as in laboratory condition it has been shown to well reproduce the signal when no absorption is present [20]. This baseline is very stable on a time basis of hours, as it is essentially determined by the increasing of the emitted power as the injection current of the laser increases during frequency scan. No effect of fringes was observed on baseline stability. During the field measurement environmental additional effect (i.e., sudden attenuation of signal from smoke or condensed vapour) could drastically change the baseline, but these cases were uncommon and were easily excluded by monitoring the χ^2 of the fit as a reliability parameter of the fit. Figure (1) shows a typical experimental absorption spectrum recorded during the present campaign, along with the best fit of Eq. (1). In the lower part it is also shown the residual in percentage between experimental signal and fitted curve, normalized with respect to the laser power. The error bar associated to each concentration value in the reported plots are determined by the propagation of the errors given by the fit for each parameters. For each fit the resulting χ^2 is monitored in order to verify the reliability of the results.

A preliminary laboratory survey has been performed on the spectrometer in order to verify its performances as well as the procedure to retrieve the gas concentrations.

Several tests were done using certified gas mixtures as reported in [19, 21] with more details: the short term reproducibility over several minutes was found to be of 0.3% and 0.5% for CO_2 and H_2O, respectively. Long term measurements (24 hours) were also performed. The dynamic range of the spectrometer was fixed in the interval 30 – 10000 ppm, as mainly limited from the maximum acceptable relative uncertainty of 10% for the retrieved concentration (lower limit), and from the maximum frequency scan to be imposed on the laser (fixed to 1 cm^{-1}) (upper limit) in order to keep the tails of the absorption within.

A comparisons between the performances of the laser spectrometer and a commercial infrared gas analyzer (IRGA) was also performed [19].

The robustness of the fitting procedure was verified with different values of the initial parameters values, as well as with Levenberg-Marquardt damping factors. Once the convergence in the fit procedure is achieved, the values of the retrieved concentration resulted not seriously biased neither by the initial values of the parameter, nor by the damping factors. Conversely the right choice of both these parameters turned out to be crucial in the number of iterations necessary to achieve convergence.

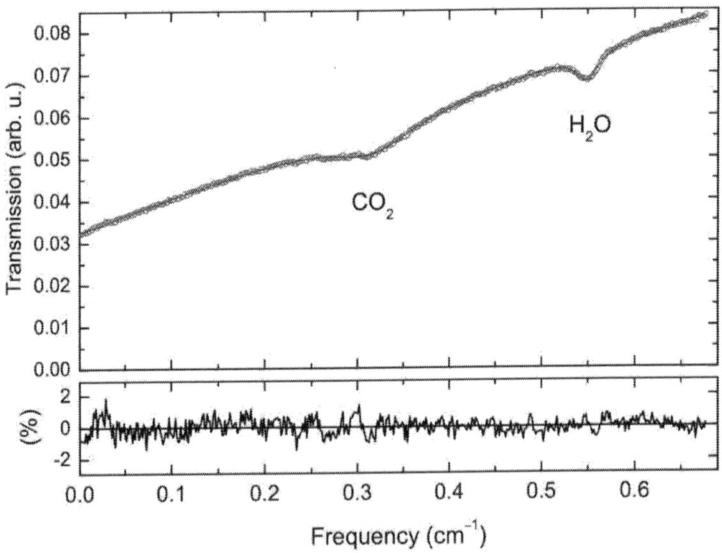

Figure 1 An example of an experimental absorption spectrum (○) and the best fit of Eq. (1) (solid line). In the plot below the residual between experimental data and the best fit is reported as percentage variation with respect to the signal. The standard deviation of the plotted residual is 0.5%.

5. FIELD MEASUREMENTS

The measurement campaign was held in August 25th–29th, 2004 at Vulcano island. During the campaign we analysed the emissions in different sites: PLB, PAL, and FOG, Fig. 2. These sites were choosen taking into account the previous studies [22–27, 29, 30, 32–37, 39–41] of the volcanic activity on Vulcano, identified as the extensive fumarole field in the northern part of La Fossa crater (100°C< T < 600°C), low-temperature fumaroles (T ≈ 100°C) in the Baia di Levante area, and widespreaded manifestations in the soil in the Vulcano Porto area and around the volcanic cone in the Palizzi valley. We performed measurements with two different configurations, with and without the breadboard cover, depending on the type of emissions we found and on the environmental conditions.

The first configuration was without the breadboard top cover, with the possibility of placing it at different heights from the ground. The second configuration was with the breadboard placed at the ground and with the top cover closed, in such a way to form an inverted accumulation chamber allowing the gas to flow from the soil to the sensing cell, through the aperture of the breadboard. The sealing of the base is ensured by the sligth burying of the breadboard in the soil. The latter was used when (1) there was no appreciable difference between the CO_2 concentration and the average background value, if measured with an open configuration, and (2) there was low water vapour emission. Indeed, when concentration of water vapour in the accumulation chamber is so high that condensation of liquid water occurs on the colder surface of the mirrors and lenses, the final power on the detector shows strong attenuation, decreasing the signal to noise ratio and finally the sensitivity. The time to determine the flux is comparable with that of other accumulation chamber. However the sampled air is measured directly at the surface were the flux occurs, with no need of move the air with pumps to the measurement cell.

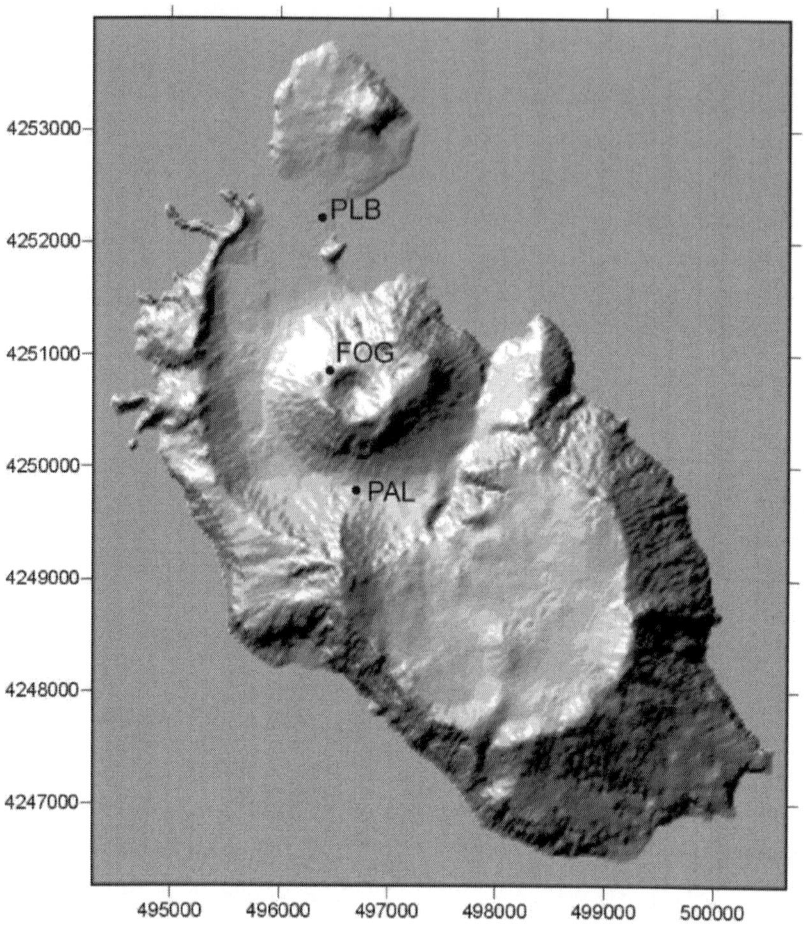

Figure 2 DTM (Digital Terrain Model) of the Vulcano island with the indication of the main measurement sites: PLB: Porto Levante Beach; FOG: Fossa Grande Crater; PAL: Valley of Palizzi.

6. GAS CONCENTRATIONS

Gas concentration measurements at Vulcano island have been performed at the three main sites (Fig. 2), namely PLB, FOG crater and PAL, and relevant data are resumed in Table 1. Different configurations were exploited.

Table 1 Mean values and standard deviations for CO_2 and H_2O concentrations and for their ratios, as measured for each site. The sites were labelled in the first column (see the text for details): measurement PLB1 to PLB3 were taken at different heigth from the ground level. All the remaining measurement were taken at ground level.

Site	T/ °C	CO_2/ppm			H_2O/ppm			CO_2/H_2O		
		Mean	(error)	st.dev.	mean	(error)	st.dev.	mean	(error)	st.dev.
Porto Levante Beach (2/8/04)										
PLB1	26	769	(19)	240	39700	(400)	5600	0.0195	(5)	0.0058
PLB2	26	477	(13)	122	36200	(400)	3500	0.0132	(4)	0.0034
PLB3	26	450	(12)	115	34500	(400)	3600	0.0131	(3)	0.0032
Palizzi (26/8/04)										
PAL1	26	384	(3)	40	25540	(80)	1140	0.0150	(1)	0.0016
PAL2	26	468	(3)	44	25830	(75)	1050	0.0182	(1)	0.0018
PAL3	24	564	(12)	120	23130	(75)	740	0.0244	(5)	0.0053

Site	T/°C	CO_2/ppm			H_2O/ppm			CO_2/H_2O		
Fossa Grande (27/8/04)										
FOG1	25	481	(24)	170	29000	(1000)	7500	0.0174	(12)	0.0090
FOG2	26	564	(15)	120	34300	(500)	3400	0.0165	(4)	0.0031
FOG3	29	541	(22)	200	35200	(2000)	19600	0.0195	(5)	0.0058
FOG4	26	505	(20)	173	30200	(900)	8000	0.0178	(3)	0.0074

Table 1 resumes the mean values and errors for CO_2, H_2O and their ratios at the different sampling points. Measurements at PLB (August 25th, 18:30 local time) have been performed setting the apparatus at various heights over the ground (PLB1 at 0 cm; PLB2 at 40 cm; PLB3 at 130 cm), in order to estimate the influence, on CO_2 and H_2O concentrations, due to the distance of the breadboard from the ground (Fig 3). Values of CO_2 concentrations range from 1500 to 450 ppm, with average values measured decreasing from 779 ppm at ground to 450 ppm at 130 cm of height. The average value is rather stable from 44 to 130 cm, indicating that we are measuring, at these heights, almost the gas concentration in the atmosphere. H_2O concentration is, on average, of about 4% at ground, slightly decreasing to 3.5% at 130 cm of height. We took also a measurement at 20 cm from the ground which however presented very noisy spectra, resulting in unreliable concentration values greater than those measured at ground level, which we cannot explain. Examples of measurements obtained at FOG (August 27th, 16:00) are shown in

Fig. 4. At this site, four different points, at distances of few meters each other, have been sampled, in order to have an idea of the variability of results within small distances. Average values for CO_2 concentrations range from 564 ppm to 481 ppm, whereas H_2O ones are between 2.9% and 3.5% as shown in Table 1. At PAL site (August 26th, 17:00), three points within distances of some meters where selected for measurements, and examples are shown in Fig. 5 while results are reported in Table 1. Average values for CO_2 concentrations lie in the range 384 to 564 ppm while for H_2O, they are between 2.3% and 2.6%.

7. FLUX MEASUREMENTS WITH ACCUMULATION CONFIGURATION

In the accumulation chamber configuration, the rate of change of the measured concentration can be modeled by the following differential equation,

$$\frac{dC_c(t)}{dt} = \alpha[C_s - C_c(t)]$$

where C_c and C_s are the concentration of the CO_2 in the chamber and in the soil, respectively, and α is the flux rate. Assuming a concentration $C_c(0)$ at the time $t = 0$ when the chamber is closed, the concentration in the chamber will change according to $C_c(t) = C_s - [C_s - C_c(0)]e^{-\alpha t}$.

By fitting the last expression to the experimental data we can retrieve the flux rate α and the concentration C_s. Examples of fitting procedure are shown in Fig. 6. The retrieved fit parameters are listed in Table 2. Knowing the volume of the chamber and the open area at base of the chamber, we can estimate the mass flux at the soil, as reported in the last column of Table 2.

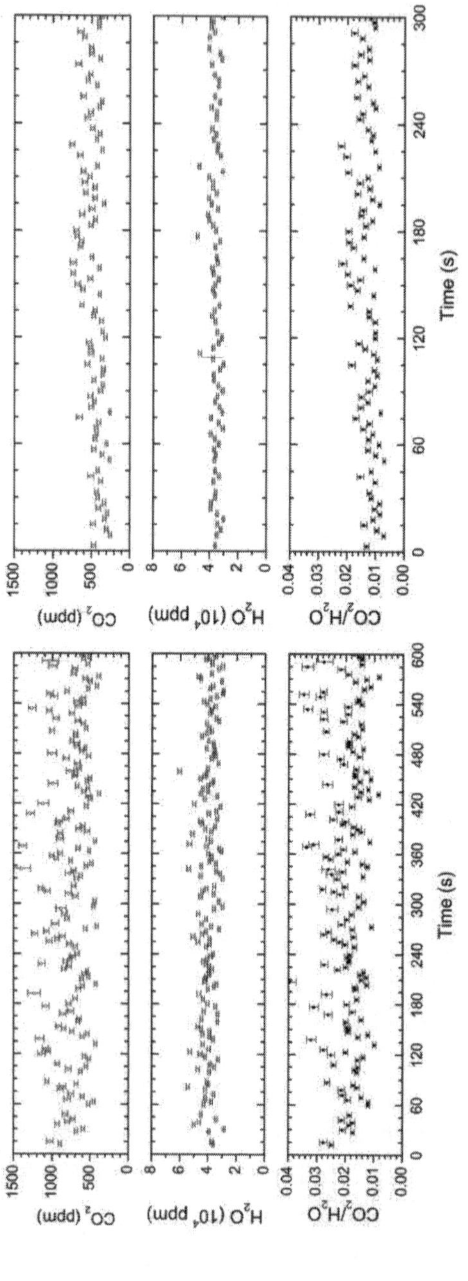

Figure 3 Two examples of concentration measurement at different heigth from the ground, 0 cm (left), 40 cm (right), made at the PLB (Aug 25th 2004) and corresponding to acquisition PLB1 and PLB3 of Table 1.

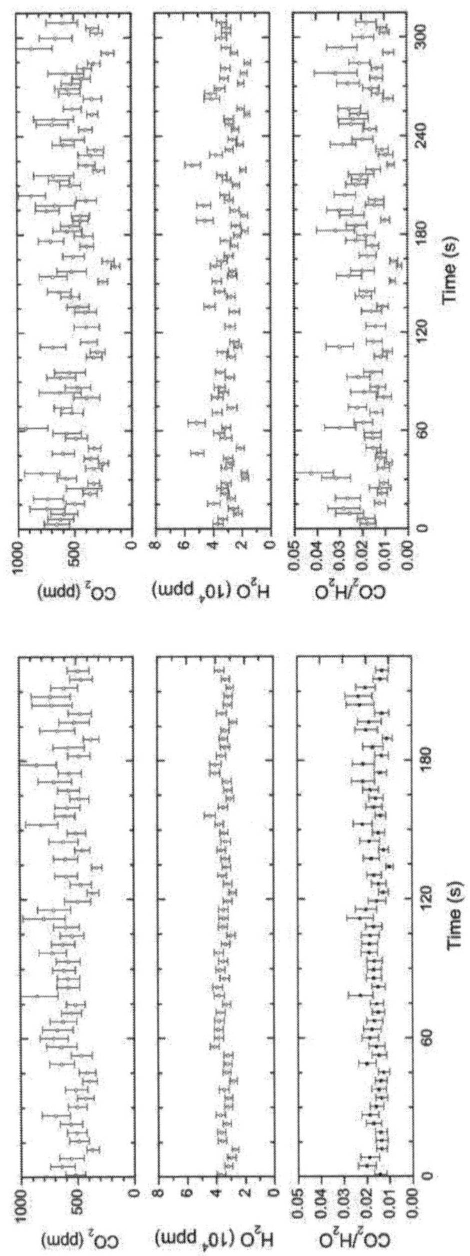

Figure 4 Two examples of measurements at FOG crater (Aug 27th 2004): left ≈ 5 meters away from the fumarole, right at the fumarole border corresponding to data FOG2 and FOG4 of Table 1, respectively.

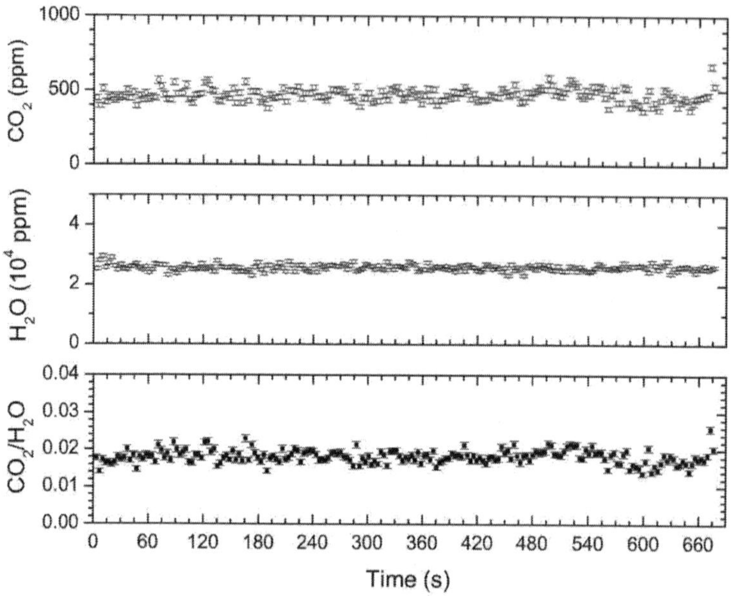

Figure 5 Concentration measurement at PAL(Aug 26th 2004). Corresponding to data PAL2 of Table 1.

Measurements of CO_2 flux at three sites of PAL (August 28th, 18:00) resulted in rates of 0.0031, 0.00090, 0.0018 s^{-1} (Table 2). At the first site, in a measurement took two days before(August 26th, 17:00), we get a rate of 0.0017 s^{-1} (Fig. 6). The concentration of water vapour remained constant during each flux measurement and its value is the same in the three different measurements of day August 28th, while for the two measurements made in the same place but in two different days (August 26th and 28th) we observed a change in water vapour concentration. This suggests that the presence of water is merely due to umidity of air. A measurements at Porto Levante Beach (August 29, 11:00)gave a flux rate of 0.0017 s^{-1}. During this measurement we observed an increase of water vapour concentration, however we believe it is likely due to the evaporation of water soil content due to sun at the measuring site after raining. Flux measurements at FOG, at distance of few meters from the main fumarole, showed not detectable flux (Table 2). Similar accumulation measurements in the fumarole failed for the condensation of water vapour on the mirrors which reduced the signal to zero.

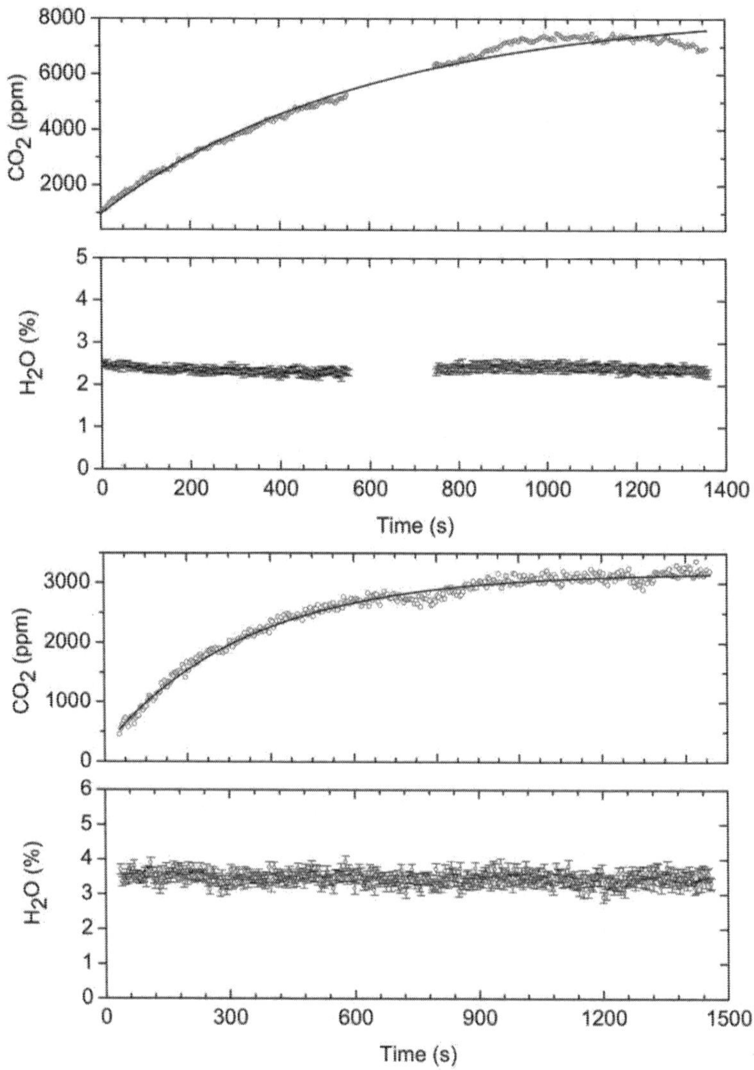

Figure 6 Flux meaurement at PAL – Aug 26th (above) and 28th (below) 2004. Solid lines represent the best fit of Eq. (3) to experimental data. Fluxes corresponding to data PAL Fl and PAL F2 of Table 2.

Table 2 Accumulation measurements. Concentration rate α and saturation values C_s retrieved by fitting Eq. (3) to the experimental data. In the last column we list the respective flux of CO_2 (at Standard Temperature and Pressure). The sites were labelled in the first column (see the text for details).

Site	T/°C	$\alpha/10^{-3}s^{-1}$	C_s/ppm	flux/g $m^{-2}s^{-1}$
Palizzi (26/8/04)				
PAL F1	24	1.65 (4)	8390 (60)	1.73
Palizzi (28/8/04)				
PAL F2	36	3.06 (6)	3140 (10)	3.20
PAL F3	40	0.90 (6)	6750 (270)	0.94
PAL F4	34	1.82 (10)	12950 (440)	1.90
Porto Levante Beach (28/8/04)				
PLB F1	33	1.72 (6)	28330 (510)	1.80

8. CONCLUSION

We have carried out a measurement campaign in Vulcano island with a portable spectrometer based on a semiconductor laser, which can monitor in-situ concentration and flux of CO_2 and H_2O. The spectrometer has a response time of the order of one second which allows the monitoring of gas dynamics on a very fast time scale. The measurements obtained at Vulcano, together with several tests, performed continuosly for several days, indicate the system has no problem to be used for long-time unattended measurements as well. The only occasional problems for a long-term remote installation could be the condensation of water vapour on the mirrors; such problem could be be reduced by heating the mirrors' surface. The average values of the CO_2/H_2O at ground level of the different measurement sites (0.0195–0.0150) is consistent with an average value measured for example at the FO fumarole using different comparative methods (FTIR, Filter pack and direct sampling) of about 0.063 [22]. Accumulation measurements have shown the presence of non-negligible CO_2 flux from the soil, at several places, which can be missed by simply monitoring the concentration in air. The measurements of gas concentration and CO_2 fluxes carried out during the present field work at Vulcano have demonstrated that laser diode spectrometers represent a suitable choice for monitoring areas with anomalous high gas concentration, in different environmental conditions. Our apparatus has shown to be suitable for continuous, in-situ monitoring of both gas concentrations and fluxes, with laboratory precision. Also, it can be considered a reliable instrument useful to the local community in areas subjected to dangerous episodes of CO_2 exhalation from soils (i.e., Ciampino and Marino towns in the Albani Hills, Central Italy [42]; Mefite d'Ansanto close to Avellino city, Southern Italy; Mammuth area Long Valley, California USA [43]; Lake Nyos and Monoun, Cameroon [44]).

ACKNOWLEDGEMENTS

We aknowledge the support of Project PON-SI.MON.A.

REFERENCES

1. Baubron J, Allard P, Sabroux J, Tedesco D, Toutain J: Soil gas emanations as precursory indicators of volcanic eruptions. J Geol Soc Lond. 1991, 148: 571-576.

2. Notsu K, Mori T, Igarashi G, Tohjima Y, Wakita H: Infrared spectral radiometer: A new tool for remote measurement of SO_2 of volcanic gas. Geochem J. 1993, 27: 361-366.

3. Nuccio P, Paonita A: Magmatic degassing of multicomponent vapors and assessment of magma depth: Application to Vulcano island (Italy). Earth Planet Sci Lett. 2001, 193: 467-481.

4. Aiuppa A, Burton M, Murè F, Inguaggiato S: Intercomparison of volcanic gas monitoring methodologies performed on Vulcano Island, Italy. Geophis Res Lett. 2004, 31: L02610-

5. Rendiconti di sorveglianza 1999–2004. [http://www.ov.ingv.it]

6. Galle B, Oppenheimer C, Geyer A, McGonigle A, Edmonds M, Horrocks L: A miniaturised ultraviolet spectrometer for remote sensing of SO_2 fluxes: a new tool for volcano surveillance. J of Volcanol Geotherm Res. 2003, 119: 241-254.

7. Elias T, Sutton AJ, Oppenheimer C, Horton KA, Garbeil H, Tsanev V, McGonigle AJS, Williams-Jones G: Comparison of COSPEC and two miniature ultraviolet spectrometer systems for SO_2 measurements using scattered sunlight. Bull Volcanol. 2006, 68: 313-322.

8. Weibring P, Swartling J, Edner H, Svanberg S, Caltabiano T, Condarelli D, Cecchi G, Pantani L: Optical monitoring of volcanic sulphur dioxide emissions – comparison between four different remote-sensing spectroscopic techniques. Opt Lasers Eng. 2002, 37: 267-284.

9. Francis P, Burton M, Oppenheimer C: Remote measurements of volcanic gas compositions by solar occultation spectroscopy. Nature. 1998, 396: 567-570.

10. Mori T, Notsu K, Tohjima Y, Wakita H, Nuccio P, Italiano F: Remote detection of fumarolic gas chemistry at Vulcano, Italy, using an FT-IR spectral radiometer. Earth Plan Sci Lett. 1995, 134: 219-224.

11. Andersen D, Farrar C: Eddy covariance measurements of CO_2 flux to the atmosphere from an area of high volcanogenic emissions, Mammoth Mountain, California. Chemical Geology. 2001, 177: 31-42.

12. Burton M, Oppenheimer C, Harrocks L, Francis P: Remote sensing of CO_2 and H_2O emission rates from Masaya volcano, Nicaragua. Geology. 2000, 28: 915-918.

13. Zimmer M, Erzinger J: Continuous H_2O, CO_2, ^{222}Rn and temperature measurements on Merapi Volcano, Indonesia. J of Volcanol Geotherm Res. 2003, 125: 25-38.

14. Faber E, Moràn C, Poggenburg J, Garzon G, Teschner M: Continuous gas monitoring at Galeras Volcano, Colombia: first evidence. J of Volcanol Geotherm Res. 2003, 125: 13-23.

15. Werle P, Slemr F, Maurer K, Kormann R, Mücke R, Jänker B: Near- and mid-infrared laser-optical sensors for gas analysis. Opt Laser Eng. 2002, 37: 101-114.

16. Richter D, Erdelyi M, Curl R, Tittel F, Oppenheimer C, Duffell H, Burton M: Field measurement of volcanic gases using tunable diode laser based mid-infrared and Fourier transform infrared spectrometers. Opt Laser Eng. 2002, 37: 171-186.

17. Gianfrani L, De Natale P, De Natale G: Remote sensing of volcanic gases with a DFB-laser-based fiber spectrometer. Appl Phys B. 2000, 70: 467-470.

18. Gianfrani L, De Natale P: Remote measurements of volcanic gases using a diode-laser-based spectrometer. Opt & Phot News. 2000, 11: 44-

19. Gagliardi G, Restieri R, De Biasio G, De Natale P, Cotrufo F, Gianfrani L: Quantitative diode laser absorption spectroscopy near 2 μm with high precision measurements of CO_2 concentration. Rev Sci Inst. 2001, 72: 4228-4233.

20. Rocco A, De Natale G, De Natale P, Gagliardi G, Gianfrani L: A diode-laser-based spectrometer for in-situ measurements of volcanic gases. Appl Phys B. 2003, 78: 235-240.

21. Gagliardi G, Restieri R, Casa G, Gianfrani L: Chemical and isotopic analysis using diode laser spectroscopy: applications to volcanic gas monitoring. Opt Lasers Eng. 2002, 37: 131-142.

22. Chiodini G, Cioni R, Marini L, Panichi C: Origin of fumarolic fluids of Vulcano Island, Italy and implications for volcanic surveillance. Bull Vulcanol. 1995, 57: 99-110.

23. Chiodini G, Cioni R, Marini L, Panichi C, Raco B, Taddeucci G: Fumarolic gases geochemistry. Acta Volcanologica. 1994, 6: 43.

24. Cardellini C, Chiodini G, Frondini F, Granieri D, Lewicki J, Peruzzi L: Accumulation chamber measurements of methane fluxeses: application to volcanic-geothermal areas and landfills. Appl Geochem. 2003, 18: 45-54.

25. Chiodini G, Frondini F, Raco B: Diffuse emission of CO_2 from the Fossa crater Vulcan island (Italy). Bull Vulcanol. 1996, 58: 41-

26. Chiodini G, Cioni R, Guidi M, Raco B, Marini L: Soil CO_2 flux measurements in volcanic and geothermal areas. Appl Geochem. 1998, 13: 543-552.

27. Aiuppa A, Federico C, Giudice G, Gurrieri S: Chemical mapping of a fumarolic field: la Fossa crater, Vulcano Island (Aeolian Islands, Italy). Geophys Res Lett. 2005, 32: L13309-

28. Allard P, Carbonelle J, Dajlevic D, le Bronec J, Morel P, Robe J, Maurenas J, Faivre-Pierret R, Martin D, Sabroux J, Zettwoog P: Eruptive and diffusive emissions of CO_2 from Mount Etna. Nature. 1991, 351: 387-391.

29. Capasso G, Inguaggiato S, Nuccio P, Peocraino G, Sortino F: Chemical changes in the gas composition of La Fossa crater fumaroloes. Acta Vulcanologica. 1993, 3: 278-282.

30. Capasso G, Favara R, Inguaggiato S: Chemical features and isotopic composition of gaseous manifestations on Vulcano Island, Aeolian Islands, Italy: An interpretative model of fluid circulation. Geochim et Cosmochim Acta. 1997, 61: 3425-3440.

31. Gurrieri S, Valenza M: Gas transport in natural porous mediums: a method for measuring CO_2 flows from the ground in volcanic and geothermal areas. Rend Soc Hal Mineral Petrologia. 1988, 43: 1151-1158.

32. Diliberto I, Guerrieri S, Valenza M: CO_2 flux from the ground. Acta Vulcanol. 1993, 3: 272-273.

33. Diliberto I, Guerrieri S: Soil CO_2 degassing in the island of Vulcano during 1992. Acta Vulcanol. 1994, 6: 35.

34. Carapezza M, Diliberto I: Helium and CO_2 soil degassing. Acta Vulcanol. 1994, 3: 273-276.

35. Carapezza M: Helium and CO_2 soil gas concentration (Vulcano). Acta Vulcanol. 1994, 6: 36-39.

36. Badalamenti B, Gurrieri S, Valenza M: Continuous monitoring (temperature, CO_2, in soil gases and reducing capacity) (Vulcano). Acta Vulcanol. 1994, 6: 46-48.

37. Carapezza M, Granieri D: CO_2 soil flux at Vulcano (Italy): comparison between active and passive methods. Appl Geochem. 2004, 19: 73-88.

38. Harvard Smithsonian Center for Astrophysics: The Hitran Database 2003. [http://www.hitran.com]

39. Nuccio P, Paonita A, Sortino F: Geochemical modeling of mixing between magmatic and hydrothermal gases: The case of Vulcano, Italy. Earth Planet Sci Lett. 1999, 167: 321-333.

40. Tedesco D, Scarsi P: Intensive gas sampling of noble gases and carbon at Vulcano Island (southern Italy). J Geophys Res. 1999, 104: 10499-10510.

41. Baubron J, Allard P, Toutain J: Diffusive volcanic emissions of carbon dioxide from Vulcano Island, Italy. Nature. 1990, 344: 51-53.

42. Pizzino L, Galli G, Mancini C, Quattrocchi F, Scarlato P: Natural gas hazard (CO_2, ^{222}Rn) within a quiescent volcanic region and its relations with tectonics: the case of Ciampino-Marino area and Albano Hill Vulcano, Italy. Natural Hazards. 2002, 27: 257-287.

43. Farrar C, Sorey M, Evans W, Howle J, Kerr B, Kennedy B, King C, Southon J: Forest-killing diffuse CO_2 emission Mammouth Mountain as a sign of magmatic unrest. Nature. 1995, 376: 675-678.

44. Kling G, Evans W, Tanyileke G, Kusakabe M, Ohba T, Yoshida Y, Hell J: Degassing Lakes Nyos and Monoun: defusing certain disaster. PNAS. 2005, 102: 14185-14190. 10.1073/ pnas. 0502274102.

8

NOVEL OPTOELECTRONIC DEVICES BASED ON SINGLE SEMICONDUCTOR NANOWIRES (NANOBELTS)

Yu Ye, Lun Dai, Lin Gan, Hu Meng, Yu Dai, Xuefeng Guo and Guogang Qin

1State Key Lab for Mesoscopic Physics and School of Physics, Peking University, Beijing 100871, China.

2College of Chemistry and Molecular Engineering, Peking University, Beijing 100871, China.

ABSTRACT

Semiconductor nanowires (NWs) or nanobelts (NBs) have attracted more and more attention due to their potential application in novel optoelectronic devices. In this review, we present our recent work on novel NB photodetectors, where a three-terminal metal–semiconductor field-effect transistor (MESFET) device structure was exploited. In contrast to the common two-terminal NB (NW) photodetectors, the MESFET-based photodetector can make a balance among overall performance parameters, which is desired for practical device applications. We also present our recent work on graphene nanoribbon/semiconductor NW (SNW) heterojunction light-emitting diodes (LEDs). Herein, by taking advantage of both graphene and SNWs, we have fabricated, for the first time, the graphene-based nano-LEDs. This achievement opens a new avenue for developing graphene-based nano-electroluminescence devices. Moreover, the novel graphene/SNW hybrid devices can also find use in other applications, such as high-sensitivity sensor and transparent flexible devices in the future.

KEYWORDS

Schottky junction, Graphene, Nanowires, Nanobelts, Optoelectronics

1. INTRODUCTION

Semiconductor single-crystalline nanowires (NWs) or nanobelts (NBs) can be grown on lattice mismatched substrates and constructed into devices with the bottom-up method on basically any substrates [1]. Hence, compared to the conventional ones, semiconductor NW- or NB-based devices have the advantage of versatility in both the material and the device structure. So far, various semiconductor NW- or NB-based nanodevices have been emerging continuously [2–4]. Developing novel high-performance nano-optoelectronic devices is not only important in diverse device applications, but also has significant meaning in exploring and realizing optoelectronic integration.

In this review, we present our research work on two types of novel optoelectronic devices based on semiconductor NWs (NBs). One is semiconductor NB metal–semiconductor field-effect transistor (MESFET)-based photodetectors [5]. In contrast to the common two-terminal single semiconductor NB (NW) photodetectors, the three-terminal NB MESFET-based photodetector can make a balance among overall performance parameters, which is desired for practical device applications. The other is novel multicolor light-emitting diodes (LEDs) based on graphene nanoribbon (GNR)/semiconductor nanowire (SNW) heterojunctions [6]. Herein,

ZnO, CdS, and CdSe NWs were employed for demonstration. At forward biases, the GNR/SNW heterojunction LEDs emitted light from ultraviolet (380 nm) to red (705 nm), which were determined by the bandgaps of the involved SNWs. This work opens a new avenue for developing diverse graphene-based optoelectronic devices [7]. These two works may help to promote nano-optoelectronic integration in the future.

2. SINGLE CDS NB MESFET PHOTODETECTOR

Photodetectors, which convert light to electric signals, are essential elements in high-resolution imaging techniques and light-wave communication, as well as in future memory storage [8]. Single NB (NW) photodetectors may find applications as binary switches, light-wave communications, and optoelectronic circuits. So far, most of the reported single NB (NW) photodetectors are two-terminal devices [3, 8, 9, 10, 11, 12, 13, 14, 15, 16, 17]. We summarize the key parameters of the three-terminal MESFET CdS NB photodetector and the reported two-terminal CdS NB (NW) photodetectors in Table 1. In general, for the two-terminal NB (NW) photodetectors, there exists a trade-off among the performance parameters, such as current responsivity (R_λ), photoresponse ratio (I_{light}/I_{dark}), and photoresponse time (rise and fall times). For example, Golberg et al. reported ohmic contact-based single CdS NB photodetectors with ultrahigh R_λ (approximately 7.3×10^4 A/W) and fast response time (approximately 20 μs of both rise and fall times); however, the I_{light}/I_{dark} was quite low (approximately 6) [15]. Wang et al. reported Schottky contact-based NW photodetectors with a higher I_{light}/I_{dark} (approximately183); however, the response time was not satisfying (approximately 320 ms of fall time) [17]. Compared to the reported two-terminal NB (NW) photodetectors, the MESFET-based photodetector can make a balance among these key parameters and have an overall improvement in the device performance.

Table 1 Comparison of the key parameters for the CdS NB (NW) photodetectors with different structures

Materials	**Structure**	**Rise; fall time**	**Responsivity (A/W)**	I_{light}/I_{dark}	**Reference**
CdS NB	Ohmic contact	746; 794 μs	approximately 38	6.0×10^3	[9]
CdS NB	Ohmic contact	approximately 20; approximately 20 μs	7.3×10^4	6.0	[15]
CdS NW	Schottky contact	-; 320 ms	-	approximately 183.0	[17]
CdS NB	MESFET	137; 379 μs	approximately 2.0×10^2	approximately 2.7×10^6	This work

A field-emission scanning electron microscopy (FESEM) image of the as-fabricated photodetector was shown in the inset of Figure 1a. Two In/Au (10:100 nm) ohmic contact electrodes were defined on one single CdS NB, while one Au (100 nm) Schottky contact electrode was defined in between the ohmic electrodes across the CdS NB. We can see that the NB has a uniform width (500 nm) along the entire length (13 μm) between the two ohmic contacts.

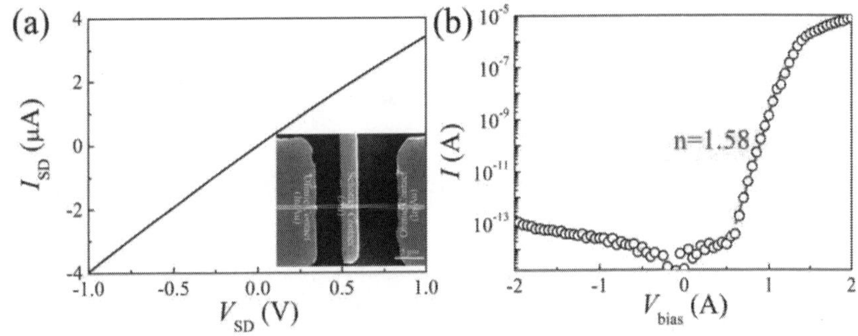

Figure 1 Typical electrical transport properties of the single CdS NB MESFETs. (**a**) The current–voltage (*I-V*) curve measured between the source and drain electrodes. Inset: a typical FESEM image of a single CdS NB MESFET-based photodetector. (**b**) The red straight line shows the fitting result with the equation $\ln(I) = \frac{qV}{nkT} + \ln(I_0)$ where I_0 is the reverse saturation current, q is the electronic charge, V is the applied bias, n is the diode ideality factor, k is the Bolzmann constant, and T is the absolute temperature.

Typical electrical transport properties of the CdS NB MESFETs are shown in Figure 1. During the electrical transport measurements, the source electrodes were grounded. Figure 1a shows the *I V* curve measured between the source and drain electrodes. It shows a linear behavior, confirming the ohmic contacts between the In/Au electrodes and the CdS NB. Figure 1b shows the *I V* curve measured between the source and gate electrodes on an exponential scale. We can see a good Schottky contact rectification behavior between the Au electrode and the CdS NB. A rectification ratio of approximately 10^8 is obtained when the voltage changes from +2 to −2 V. The turn-on voltage is around 1.25 V. The typical transfer characteristic curves of the CdS NB MESFETs with and without light illumination are depicted in Figure 2a. We can see that the V_{th} shifts from −2.9 to −3.8 V when the light is switched from off-state to on-state. This phenomenon can be understood as follows: there are two processes involved when the as-fabricated device is upon above-bandgap illumination. One is that the channel conductance increases due to the photon-generated electrons and holes; the other is that photon-generated electrons and holes at the Schottky junction are separated by the strong local electric field [16, 18], which may reduce the electron–hole recombination rates and lower the barrier height [16–19]. Both processes will make the CdS NB channel more difficult to be depleted, and hence, the V_{th} shifts to a more negative value. The photocurrent response of a control device, a two-terminal CdS NB photodetector, is shown in Figure 2b. It has a small photoresponse ratio (I_{light}/I_{dark} approximately 2.78) and a long decay tail (tens of seconds). The photocurrent response of the CdS NB MESFET measured at $V_G = 0$ V is shown in Figure 2c. We can see that the average dark current (light-off) and photocurrent (light-on) are about 1.77 and 1.86 μA, respectively, resulting in an I_{light}/I_{dark} of approximately 1.05. Again, a long decay tail of tens of seconds can be observed.

Figure 2d shows the on/off photocurrent response of the CdS NB MESFET-based photodetector measured at $V_G = -3.8$ V, which is the threshold voltage of the MESFET under light illumination. We can see that the average dark current and photocurrent are about 26 fA and 70 nA, respectively, resulting in a I_{light}/I_{dark} as high as approximately 2.7×10^6. To the best of our knowledge, this is so far among the highest reported values for single NB (NW) photodetectors [3, 8, 9, 10, 11, 12, 13, 14, 15, 16, 17]. In addition, the photoresponse processes (both rise and decay processes) are quite fast, which have exceeded the detection limit (0.3 s) of the measurement apparatus (Keithley 4200, Cleveland, OH, USA).

The R_λ, defined as the photocurrent generated per unit power of incident light on the effective illuminated area of a photoconductor, and the external quantum efficiency (EQE), defined as the number of electrons detected per incident photon, are two critical parameters for photodetectors. The R_λ and EQE can be calculated with equations $R_\lambda = \frac{\Delta I}{P_\lambda S}$ and $\mathrm{EQE} = \frac{hcR_\lambda}{e\lambda}$ [11], respectively. Here, ΔI is the difference between the photocurrent and the dark current, P_λ is the light power density, S is the effective illuminated area, h is Planck's constant, c is the velocity of light, e is the electronic charge, and λ is the light wavelength. Using $\Delta I = 7.0 \times 10^{-8}$ A (measured from Figure 2d), $P_\lambda = 5.3$ mW/cm^2$S = 500$ nm × 13 μm (measured from the inset of Figure 1a),$\lambda = 488$ nm, the R_λ and EQE of the CdS NB MESFET photodetector can be estimated to be approximately 2.0×10^2 A/W and 5.2×10^2, respectively.

In order to further investigate the detailed photoresponse times of the single CdS NB MESFET photodetector, we employed a 200-MHz digital oscilloscope (Tektronix DPO2024, Beaverton, OR, USA) with a 10-MΩ impedance and an optical chopper working at a frequency of 1,000 Hz, as shown in Figure 2e. From the close-up of the measured result shown in Figure 2f, the rise timer, defined as the time needed for the photocurrent to increase from 10 % i_{peak} to 90 % i_{peak}, is 137 μs and the decay timed, defined analogously, is 379 μs.

We attribute the overall high performance of our CdS NB MESFET-based photodetectors to the unique advantage of the MESFET structure. Compared to two-terminal photodetectors, there are two main advantages of the MESFET-based photodetectors. First, it has a much lower dark current because the applied negative gate voltage (in our case, the threshold voltage under illumination) helps to deplete the channel carriers. Second, this gate depletion effect will also cause a fast current recovery when the light is turned off. Consequently, the decay tail, which is normally observed in a two-terminal photodetectors, is suppressed in the MESFET-based photodetectors.

3. MULTICOLOR GNR/SNW HETEROJUNCTION LEDS

By taking advantage of both graphene and SNWs, we have fabricated, for the first time, the graphene-based nano-LED [6]. This achievement opens a new avenue for developing graphene-based nano-electroluminescence devices. Moreover, the novel graphene/SNW hybrid devices can also find use in other applications, such as high-sensitivity sensor and transparent flexible devices in the future.

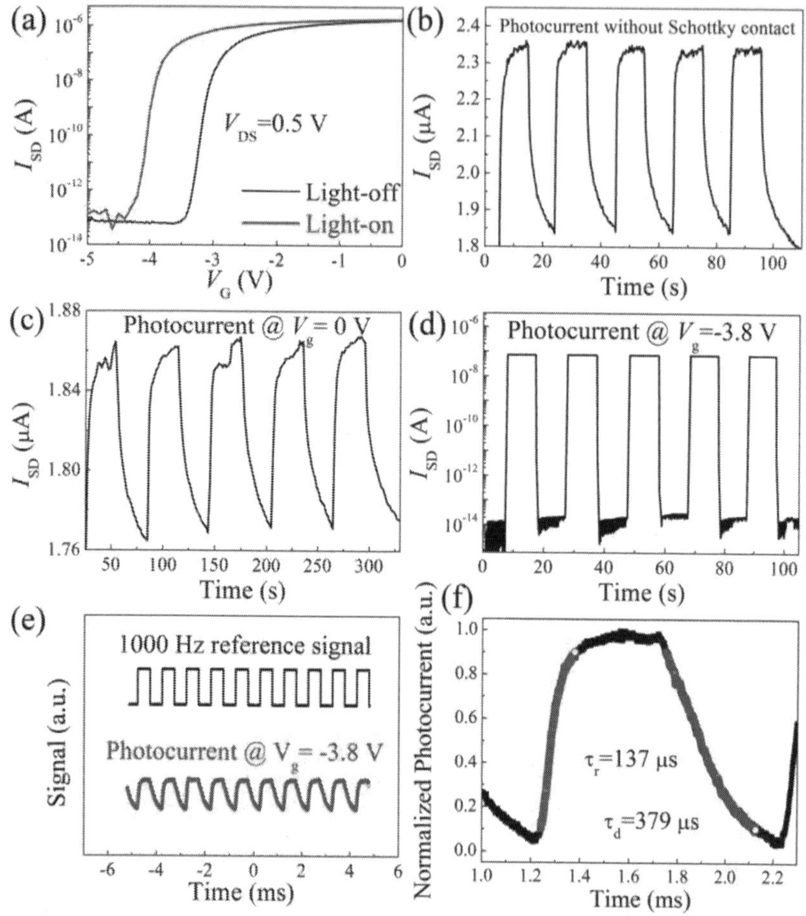

Figure 2 The typical light response properties of the single CdS NB MESFET-based photodetectors. (**a**) The transfer characteristics of a CdS NB MESFET-based photodetector measured in the dark (black line) and under illumination (red line). (**b**) On/off photocurrent response of the CdS NB without Schottky contact as a function of time. (**c**) On/off photocurrent response of the CdS NB MESFET-based photodetector with $V_G = 0$ V as a function of time on a linear scale. (**d**) On/off photocurrent response of the CdS NB MESFET-based photodetector with $V_G = -3.8$ V as a function of time on an exponential scale. (**e**) A transient response of the CdS NB MESFET-based photodetector ($V_G = -3.8$ V, $V_{DS} = 0.5$ V) along with a reference signal of the chopped light with a frequency of 1,000 Hz. (**f**) A close-up of the result shown in (e).

Both the n-type NWs [20–22] and the graphene [23] used in this work were synthesized via the CVD method. Before device fabrication, the graphenes were transferred by the stamp method with the help of polymethyl methacrylate [24] to Si/300-nm SiO_2substrates for Raman and electrical property characterizations, to quartz substrates for transparency characterization, and transferred to carbon-coated grids for high-resolution transmission electron microscopy (HRTEM) characterization (Tecnai F30, FEI, Eindhoven, The Netherlands). Their electrical properties were measured by a Hall effect measurement system (Accent HL5500, York, England).

The HRTEM, Raman, and transparency characterization results for the as-synthesized graphenes (Figure 3) demonstrate that the graphenes have high quality, monolayer, and high transparency. The typical sheet resistance, hole concentration, and hole mobility of the graphenes are about 345 Ω/sq, 1.84×10^{14} cm^{-2}, and 98.6 $cm^2/V{\cdot}s$, respectively.

The fabrication processes of a GNR/SNW heterojunction LED are shown in Figure 4. Figure 5a shows an FESEM image of an as-fabricated GNR/CdS NW heterojunction LED. The *I V* curve (Figure 5b) of the LED shows an excellent rectification characteristic. An on/off current ratio of approximately 3.4×10^7 can be obtained when the voltage changes from +1.5 to −1.5 V. The turn-on voltage is around 1.1 V. In view of the high conductivity and near-zero bandgap characteristics of the GNR [25], the heterojunction structure of the GNR/CdS NW can be considered approximately as a metal–semiconductor contact of the Schottky model [26]. We can deduce that the diode ideality factor $n = 1.58$. Note that the GNR/ZnO NW and GNR/CdSe NW heterojunctions show similar rectification characteristics as described above, with the turn-on voltages to be about 0.7 and 1.2 V, respectively.

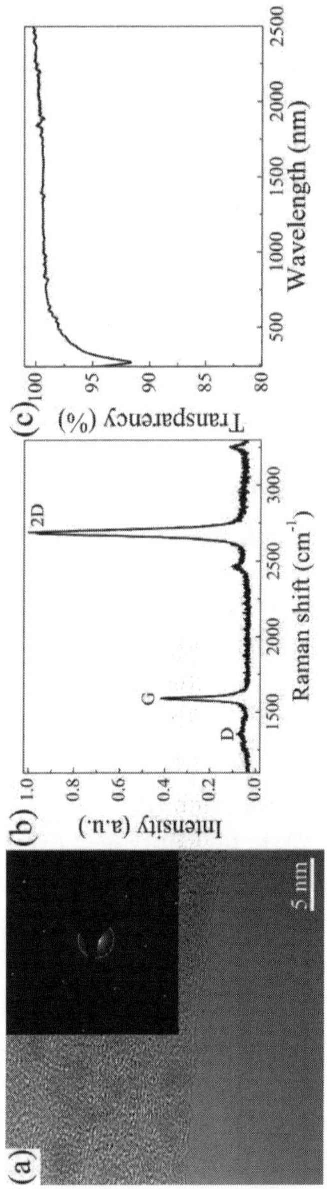

Figure 3 Properties of as-synthesized graphene. (**a**) Typical HRTEM image of an as-synthesized graphene, indicating the formation of monolayer graphene. Inset: selected area electron diffraction pattern of the graphene. (**b**) Raman spectrum of an as-synthesized graphene on a Si/300-nm SiO_2 substrate. (**c**) The transparency spectrum of the graphene on a quartz substrate.

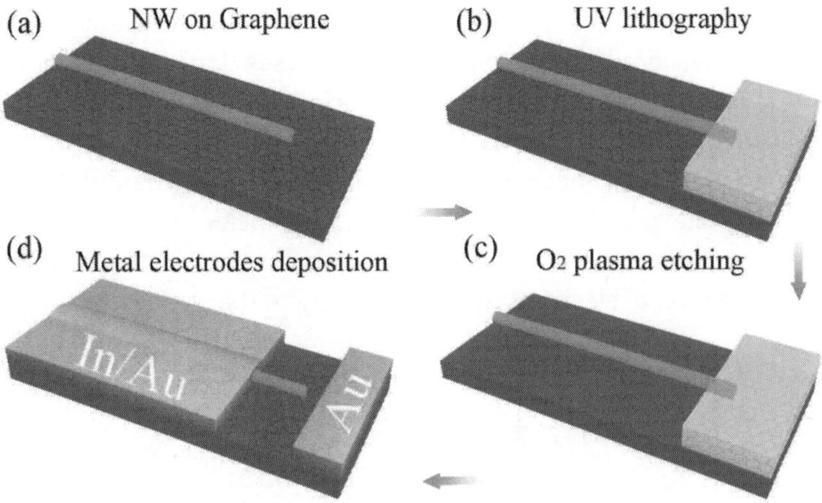

Figure 4 Schematic illustration of the fabrication processes of a GNR/SNW heterojunction LED. (**a**) The as-synthesized large-scale graphene was transferred to a Si/SiO_2 substrate. After that, SNW suspension was dropped on the graphene. (**b**) A photoresist pad was patterned to cover one end of a SNW by UV lithography and development processes. (**c**) Oxygen plasma etching was used to remove the exposed graphene. After that, the GNR formed under the SNW. (**d**) After removing the photoresist, In/Au and Au ohmic contact electrodes to SNW and graphene pad were defined, respectively. It is worth noting that because an undercut was formed during the oxygen plasma etching process (Figure 4c) [6, 27], the In/Au electrode on the SNW will not contact with the GNR beneath.

Figure 6a,b,c shows the electroluminescence (EL) images (Olympus BX51M, Shinjuku-ku, Japan) of the GNR/SNW (ZnO, CdS, CdSe, respectively) heterojunction LEDs at a forward bias of 5 V. Except for the ZnO NW case (where the emitting light is invisible ultraviolet light) in Figure 6a, strong emitting light spots can be seen clearly with naked eyes at the exposed ends of the NWs. For the CdS NW case (Figure 6b), we can see another glaring light spot on the NW. This may be due to the scattering from the defect or adhered particle on the CdS NW [28]. Figure 6d,e,f shows the room-temperature EL spectra measured at various forward biases for the GNR/SNW heterojunction LEDs, where the SNWs are ZnO, CdS, and CdSe NWs, respectively. For all the LEDs, EL intensities increase with the forward biases. The peak wavelength of each EL spectrum (380, 513, and 705 nm, respectively, from (d) to (f)) coincides with the band-edge emission of the SNW involved. This indicates that the radiative recombination of electrons and holes occurs in the SNWs.

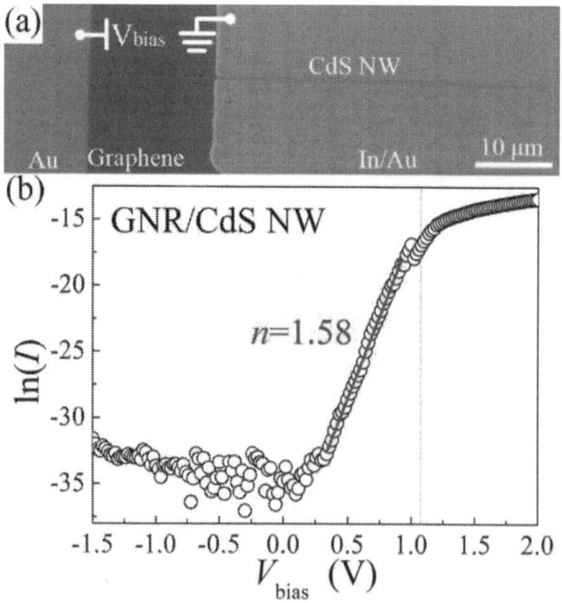

Figure 5 FESEM image and room-temperature *I* - *V* characteristic of as-fabricated GNR/CdS NW heterojunction LED. (**a**) FESEM image of an as-fabricated GNR/CdS NW heterojunction LED. (**b**) Room-temperature *I-V* characteristic of the LED in (a) on a semilog scale. The red straight line shows the fitting result of the *I-V* curve by the equation $\ln(I) = \frac{qV}{nkT} + \ln(I_0)$.

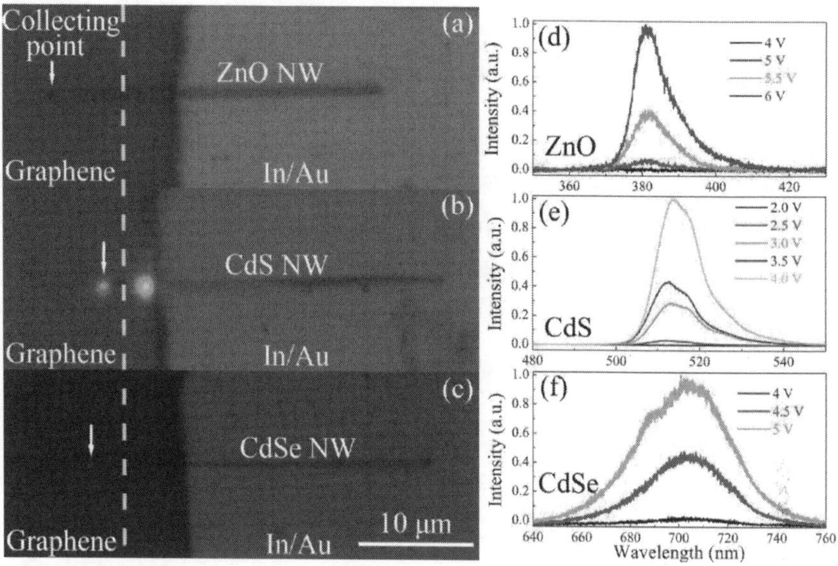

Figure 6 Optical images and room-temperature EL spectra of the GNR/SNW heterojunction LEDs. (**a, b, c**) The optical images of the GNR/SNW (ZnO, CdS, CdSe, respectively) heterojunction LEDs at a forward bias of 5 V. Dashed lines were used to demarcate the graphenes from the substrates. White arrows: the light collecting points during the EL measurements. (**d, e, f**) Room-temperature EL spectra for GNR/SNW (ZnO, CdS, CdSe, respectively) heterojunction LEDs at various forward biases.

We can qualitatively understand the mechanism of the light emitting for the GNR/SNW heterojunction LEDs by studying the energy band diagrams. Figure 7a shows the thermal equilibrium energy band diagram of a graphene/n-type semiconductor structure, where the work function of graphene is Φ, and the electron affinity of the semiconductor is χ. E_g and E_F correspond, respectively, to the bandgap and the Fermi level of the semiconductor. It is worth noting that because the graphene used in this work has a very high conductivity and can be taken as a metal, the graphene/SNW heterostructure herein can be taken as a kind of Schottky junction. At the thermal equilibrium contacting state, the energy band of the semiconductor will bend upward at the graphene/semiconductor interface due to the difference between their work functions, and the Fermi levels at the two sides are brought into coincidence. Under a forward bias (i.e., a positive bias on graphene), the built-in potential is lowered. Therefore, more electrons will flow from n-type semiconductor to graphene, and simultaneously, more holes will flow from graphene to n-type semiconductor. Herein, the injected holes have a higher radiative recombination with the electrons in the SNW (the direct bandgap semiconductor). Accordingly, the EL spectra are determined mainly by the band-edge emission of the SNWs.

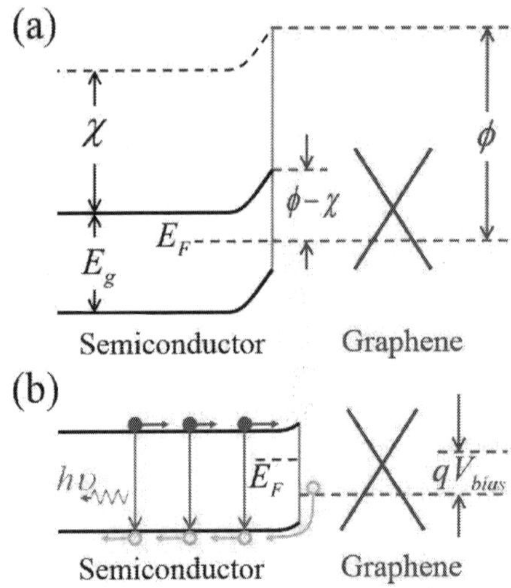

Figure 7 Schematic illustration of the energy band diagrams of a graphene/semiconductor heterojunction. (**a**) The thermal equilibrium energy band diagram. (**b**) The energy band diagram of the heterojunction under a forward bias. Φ: the work function of graphene; χ: the electron affinity of the semiconductor.

It is worth noting that the GNR/NW structure has clear advantage over the conventional Schottky structure. For comparison, we have fabricated various metal/SNW Schottky structures, where the NWs used are identical to those reported in this work. Unfortunately, no EL can be observed in these structures. We attribute this to the well-known luminescence quenching effect caused by the involved metal [29]. Moreover, in our face-to-face contact LED, the active region, where the radiative recombination occurs, is larger and the series resistance is smaller, compared to the crossed NWs or NW/Si pad heterojunction structures [20,30, 31]. These merits may benefit high-efficiency EL and even electrically driven laser in the future.

4. CONCLUSION

We review two types of novel nano-optoelectronic devices developed in our group recently. One is the photodetector, which converts light to electric signals. Our MESFET-based photodetectors have ultrahigh I_{light}/I_{dark} (approximately 2.7×10^6) and fast response (rise time, approximately 137 μs; decay time, approximately 379 μs) simultaneously. The other is LED, which converts electric power to light. At forward biases, our novel GNR/SNW heterojunction LEDs emitted light with wavelengths varying from ultraviolet (380 nm) to red (705 nm), which were determined by the bandgaps of the involved SNWs. These two types of nano-optoelectronic devices may find diverse applications in future nano-optoelectronic integration.

ACKNOWLEDGMENT

This work was supported by the National Natural Science Foundation of China (nos. 61125402, 51172004, 11074006, 10874011, 50732001), the National Basic Research Program of China (nos. 2012CB932703, 2007CB613402), and the Fundamental Research Funds for the Central Universities.

COMPETING INTERESTS

The authors declare that they have no competing interests.

AUTHORS' CONTRIBUTIONS

YY carried out the device fabrications, participated in the statistical measurements, and drafted the manuscript. LD and GQ participated in the instruction, discussion, and manuscript revision. LG and XG synthesized the graphene. HM participated in the device design. YD synthesized the CdSe NWs. All authors have read and approved the final manuscript.

REFERENCES

1. Yang PD, Yan RX, Fardy M: Semiconductor nanowire: what's next? *Nano Lett* 2010, 10: 1529.

2. Duan XF, Huang Y, Agarwal R, Lieber CM: Single-nanowire electrically driven lasers. *Nature* 2003, 421: 241.

3. Shen GZ, Chen D: One-dimensional nanostructures for photodetectors. *Recent Pat Nanotechnol* 2010, 4: 20.

4. Zhai TY, Fang XS, Li L, Bando Y, Golberg D: One-dimensional CdS nanostructures: synthesis, properties, and applications. *Nanoscale* 2010, 2: 168.

5. Ye Y, Dai L, Wen XN, Wu PC, Pen RM, Qin GG: High-performance single CdS nanobelt metal–semiconductor field-effect transistor-based photodetectors. *ACS Appl Mater Interfaces* 2010, 2: 2724.

6. Ye Y, Gan L, Dai L, Meng H, Wei F, Dai Y, Shi ZJ, Yu B, Guo XF, Qin GG: Multicolor graphene nanoribbon/semiconductor nanowire

heterojunction light-emitting diodes. *J Mater Chem* 2011, 21: 11760.

7. Bonaccorso F, Sun Z, Hasan T, Ferrari AC: Graphene photonics and optoelectronics. *Nat Photon* 2010, 4: 611.

8. Kind H, Yan HQ, Messer B, Law M, Yang PD: Nanowire ultraviolet photodetectors and optical switches. *Adv Mater* 2002, 14: 158.

9. Jie JS, Zhang WJ, Jiang Y, Meng XM, Li YQ, Lee ST: Photoconductive characteristics of single-crystal CdS nanoribbons. *Nano Lett* 1887, 2006: 6.

10. Gao T, Li QH, Wang TH: CdS nanobelts as photoconductors. *Appl Phys Lett* 2005, 86: 173105.

11. Zhai TY, Fang XS, Liao MY, Xu XJ, Li L, Liu BD, Koide Y, Ma Y, Yao JN, Bando Y, Golberg D: Fabrication of high-quality In2Se3 nanowire arrays toward high-performance visible-light photodetectors. *ACS Nano* 2010, 4: 1596.

12. Fang XS, Xiong SL, Zhai TY, Bando Y, Liao MY, Gautam UK, Koide Y, Zhang XG, Qian YT, Golberg D: High-performance blue/ultraviolet-light-sensitive ZnSe-nanobelt photodetectors. *Adv Mater* 2009, 21: 5016.

13. Fang XS, Bando Y, Liao MY, Gautam UK, Zhi CY, Dierre B, Liu BD, Zhai TY, Sekiguchi T, Koide Y, Golberg D: Single-crystalline ZnS nanobelts as ultraviolet-light sensors. *Adv Mater* 2034, 2009: 21.

14. Zhai TY, Liu HM, Li HQ, Fang XS, Liao MY, Li L, Zhou HS, Koide Y, Bando Y, Golberg D: Centimeter-long V2O5 nanowires: from synthesis to field-emission, electrochemical, electrical transport, and photoconductive properties. *Adv Mater* 2010, 22: 2547.

15. Li L, Wu PC, Fang XS, Zhai TY, Dai L, Liao MY, Koide Y, Wang HQ, Bando Y, Golberg D: Single-crystalline CdS nanobelts for

excellent field-emitters and ultrahigh quantum-efficiency photodetectors. *Adv Mater* 2010, 22: 3161.

16. Zhou J, Gu YD, Hu YF, Mai WJ, Yeh PH, Bao G, Sood AK, Polla DL, Wang ZL: Gigantic enhancement in response and reset time of ZnO UV nanosensor by utilizing Schottky contact and surface functionalization. *Appl Phys Lett* 2009, 94: 191103.

17. Wei TY, Huang CT, Hansen BJ, Lin YF, Chen LJ, Lu SY, Wang ZL: Large enhancement in photon detection sensitivity via Schottky-gated CdS nanowire nanosensors. *Appl Phys Lett* 2010, 96: 013508.

18. Ye Y, Dai L, Wu PC, Liu C, Sun T, Ma RM, Qin GG: Schottky junction photovoltaic devices based on CdS single nanobelts.*Nanotechonology* 2009, 20: 375202.

19. Jin YZ, Wang JP, Sun BQ, Blakesley JC, Greenham NC: Solution-processed ultraviolet photodetectors based on colloidal ZnO nanoparticles. *Nano Lett* 2008, 8: 1649.

20. Yang WQ, Huo HB, Dai L, Ma RM, Liu SF, Ran GZ, Shen B, Lin CL, Qin GG: Electrical transport and electroluminescence properties of n-ZnO single nanowires. *Nanotechnology* 2006, 17: 4868.

21. Ye Y, Dai Y, Dai L, Shi ZJ, Liu N, Wang F, Fu L, Peng RM, Wen XN, Chen ZJ, Liu ZF, Qin GG: High-performance single CdS nanowire (nanobelt) Schottky junction solar cells with Au/graphene Schottky electrodes. *ACS Appl Mater Interfaces* 2010, 2: 3406.

22. Ye Y, Ma YG, Yue S, Dai L, Meng H, Li Z, Tong LM, Qin GG: Lasing of CdSe/SiO2 nanocables synthesized by the facile chemical vapor deposition method. *Nanoscale* 2011, 3: 3072.

23. Gan L, Liu S, Li DN, Gu H, Cao Y, Shen Q, Wang ZX, Wang Q, Guo XF: Facile fabrication of the crossed nanotube-graphene junctions.*Acta Phys – Chim Sin* 2010, 26: 1151.

24. Reina A, Jia XT, Ho J, Nezich D, Son H, Bulovic V, Dresselhaus MS, Kong J: Large area, few-layer graphene films on arbitrary substrates by chemical vapor deposition. *Nano Lett* 2010, 4: 2689.

25. Castro Neto AH, Guinea F, Peres NMR, Novoselov KS, Geim AK: The electronic properties of graphene. *Rev Mod Phys* 2009, 81: 109.

26. Thomas D, Boettcher J, Burghard M, Kern K: Photocurrent distribution in graphene-CdS nanowire devices. *Small* 1868, 2010: 6.

27. Liu C, Dai L, Ye Y, Sun T, Peng RM, Wen XN, Wu PC, Qin GG: High-efficiency color tunable n-CdSxSe1-x/p+-Si parallel-nanobelts heterojunction light-emitting diodes. *J Mater Chem* 2010, 20: 5011.

28. Ma RM, Wei XL, Dai L, Liu SF, Chen T, Yue S, Li Z, Chen Q, Qin GG: Light coupling and modulation in coupled nanowire ring-Fabry-Pérot cavity. *Nano Lett* 2009, 9: 2679.

29. Flynn RA, Kim CS, Vurgaftman I, Kim M, Meyer JR, Mäkinen AJ, Bussmann K, Cheng L, Choa FS, Long JP: A room-temperature semiconductor spaser operating near 1.5 μm. *Opt Express* 2011, 19: 8954.

30. Zhong ZH, Qian F, Wang DL, Lieber CM: Synthesis of p-type gallium nitride nanowires for electronic and photonic nanodevices. *Nano Lett*2003, 3: 343.

31. Gudiksen MS, Lauhon LJ, Wang JF, Smith DC, Lieber CM: Growth of nanowire superlattice structures for nanoscale photonics and electronics. *Nature* 2002, 415: 617.

9

THE USE OF LASER IN GARMENT MANUFACTURING: AN OVERVIEW

Rajkishore Nayak, Rajiv Padhye

School of Fashion and Textiles, RMIT University, 25 Dawson St., Brunswick 3056, Australia

ABSTRACT

Laser is being used in apparel industry from nineteenth century for various garment manufacturing applications. There are several advantages of using laser over the conventional processes in cutting, engraving, embossing, denim fading and other applications. In addition, product damage potential is reduced, no/less consumables are needed and no problem of toxic by-product disposal as found in some processes. Today's laser equipment is a result of continuous research and development of earlier products, which has undergone several changes. The initial laser systems were cumbersome, hard to run and difficult to maintain. However, the modern laser systems are simpler in operation and maintenance. Furthermore, the earlier systems were involved with more safety issues and needed the gasses to be constantly replenished. The garment manufactures around the globe should take the advantage of laser application in the post multi-fibre agreement regime to make their products more competitive. This review focuses on the technology of laser including various classifications. In addition it includes the applications of laser in garment manufacturing, their potential hazards and health related concerns.

KEYWORDS

Laser, Fabric cutting, Denim fading, Mass customization, Engraving

1. INTRODUCTION

Laser is being used in apparel industry from ninenteenth century. Recently the use of laser in apparel industry is increasing in cutting garment patterns, patterning designer neckties, 3D body scanning, denim fading and engraving leather (Nayak and Khandual 2010; Istook and Hwang 2001; Simmons and Istook 2003; Ortiz-Morales et al. 2003; Ozguney 2007; Bahtiyari 2011). The major reasons for wide application of laser in garment industries may be due to reduced cost, flexibility and anti-counterfeiting (Kovacs et al. 2006; Tarhan and Sarıışık, 2009; Yuan et al.2012). For example the artwork of high-end necktie producers are digitally stored rather than physical patterns to lower the theft risk. When needed, the digital patterns are converted into physical samples using lasers (Lucas et al. 2015; Kan2015). Recently, the application of laser in denim engraving is increasing rapidly for value addition by replacing the traditional denim-distressing technics, which will take the denim segment to a height of sophistication that can never be realised by non-laser methods (Kan 2014a). The unique nature of the garment manufacturing industry needs laser applications, which combines performance with reduced cost by eliminating the handling systems used in non-laser workstations.

Laser is an energy source, whose intensity and power can be precisely controlled. The laser beam can be focused to a desired object at specific angle depending on the application. Laser can help to cut a variety of objects ranging from flexible fabric to rigid and strong metal (Belli et al. 2005; Ondogan et al. 2005). Laser equipment are becoming widely popular in textile, leather and garment industries due to the advantage of accuracy, efficiency, simplicity and the scope of automation (Kan et al. 2010; Lu et al. 2010; Sutcliffe et al. 2000). For example the conventional cutting tools such as band blades, discs and reciprocating knives suffer from the limitations especially on delicate materials as the cutting force can displace the material, which can lead to inaccurate cutting (Nayak and Padhye2015b). The traditional cutting methods often require an operator with full attention (Vilumsone-Nemes 2012). Hence, there is a trade-off between the maximum speed of cutting and the accuracy. In addition, other limitations include intricacy of the cut components, tool longevity and machine downtime during tool servicing. These limitations are not present in laser devices, which helps to achieve improved efficiency and reduced cost.

Laser cutting with processing speed, high precision, simple operation and other advantages, so in most industries can be used to, but it is in the clothing industry, leather processing plays a different role, can quickly cut leather graphics and draw precise clothing model (Potluri and Atkinson 2003; Ready et al. 2001). The benefit of laser cutting operations involve highly collimated beam that can be focused to a very fine dot of extremely high energy density for precise cutting. Garment industry pay attention to the size of the garment when processing precision, the purpose is to achieve high efficiency, exquisite tailoring, it is better than the traditional manual cutting by spectrum.

Although there are several applications of laser in apparel manufacturing, there are limited publications on this. Hence, an attempt was made in the current article to perform a thorough review that covers various applications and recent developments in laser that can provide guidelines for future directions. In addition, a systematic description has been done on the technology of laser including various types of laser. In addition, this review includes the applications of laser in garment manufacturing, their potential hazards and health related concerns.

2. LASER TECHNOLOGY

LASER is *light amplification by stimulated emission of radiation.* Laser is an electromagnetic radiation, produced by the atoms due to energy states are changed in some materials (Dowden 2009). The atoms promoted to higher energy states emit laser in the form of light by the process known as "stimulated emission". Subsequently, this laser is being amplified in a suitable lasing medium with the help of mirrors. The final laser delivers from the equipment as a stream similar to light. The colour of the laser depends on its wavelength. The most widely used unit used to express the wavelength of a laser is in nanometre (nm).

Laser light (Fig. 1) emitted from a laser has four fundamental characteristics: Intensity, Coherency, Monochromaticity, and Collimation, which distinguish it from natural light. A high energy concentration per unit area of the beam is present in the laser. A laser beam can be of very high intensity with 1–2 mm of beam diameter and an output power of some milliwatts (mW). All the lasers are not powerful in spite of the light with high intensity as intensity is estimated from the power output/area.

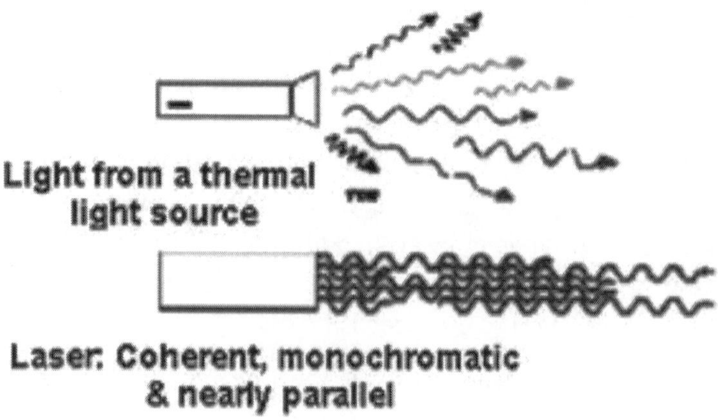

Figure. 1 An ordinary light and a Laser beam

From Fig. 1 it can be observed that the ordinary light (from a thermal source) is incoherent, i.e., light waves are generated at different times and propagate in all possible directions randomly. However, laser is coherent as shown in the figure due to the waves are in phase while they propagate. Ordinary light is composed of all the colours in the visible region, but laser light is of a single colour or monochromatic. The coherent nature in addition to the monochromaticity results in highly collimated laser. As all the waves propagate in same phase in parallel lines, there is almost no divergence as observed in light. This property of laser helps in achieving high intensity even after travelling a long distance. The energy concentration of the beam can be increased by manifold when the beam is focused as a point with the help of optical lens.

For the production of the laser beam a lasing material is stimulated with electric discharge in an enclosed container. A number of partial mirrors help in multiple internal reflections that lead to obtain the desired coherency and power output. Mirrors or other optical devices help in focusing the laser to a lens, which then reflects the laser to the working zone. In order to enable the cutting action by a laser away from the edge, it should be pierced before the laser cut using a high-power pulsed laser beam.

2.1 Classes of lasing media

A substance, which can be excited by input energy, is called a lasing media. The lasing media have the ability to stay in a metastable state. It is generally transparent to light. The substance used as lasing media can be a solid, gas, liquid dye or semiconductor. Solid lasing media is used for producing solid-state laser. Due to the high density of lasing atoms, solid-state lasers can generate higher power output per unit volume compared to the gaseous-state lasers. A single gas or gaseous mixture can be used to produce gaseous-state lasers. Gaseous-state lasers are produced by passing the electric discharge in the gaseous lasing medium. The most common type of gaseous laser is produced from helium–neon (He–Ne). This laser is mainly used in teaching laboratories, at construction sites and supermarkets.

Liquid dyes can also produce laser by using ultraviolet (UV) light. These liquid dyes can be prepared by dissolving the solid dye particles in suitable solvent. The dyes and solvents with similar energy levels are selected, that help to form a continuum for achieving high energy output with wavelengths of visible light. Better tunability is the unique feature of a dye laser, which distinguishes it from the solid laser. A single dye produces a single wavelength whereas dyes can be mixed to obtain a range of wavelengths. Semiconductors can also be used to produce laser with the help of electric discharge. These lasers are characterised by their efficiency, miniature size and durability.

2.2 Laser classification based on hazards

Since the early 1970s, lasers were being classified into four classes and some subclasses depending on their wavelength and maximum output power. This classification was based on the severity of the damage to a person when exposed to laser. These classes can be from 1 to 4. A Class 1 laser is absolutely not dangerous when used, whereas the Class 4 laser is the most dangerous. The existing system was the revised system since 2002, prior to that the old system was used. In the new system amendments have been done to certain types of lasers having a lower hazard than mentioned in the old system. Based on the output power in a specific wavelength, lasers are classified. It is essential the correct information on the laser class, potential hazards and safety instructions are specified by the equipment manufacturers. In the classes 1–4 laser, there are the subclasses 1M, 2M, 3A and 3B. The new system currently in use (shown in Table 1) uses Arabic numerals (1–4), whereas it was classified with Roman numerals (I–IV) in the old system.

Table 1 The new system for laser classification

Laser class	Features
Class 1	The Class 1 laser is the safest under normal use. These lasers may pose a risk when viewed with a telescope or microscope of sufficiently high aperture
Class 1M	The Class 1M laser is safe during normal use. However, when passed through a magnifying device such as a microscope can pose hazard. A laser falls in this class if the power that can pass through the pupil of a naked eye is lower than the accessible emission level (AEL) for Class 1
Class 2	A Class 2 laser is the visible-light laser (400–700 nm). It is safe as the blink reflex will limit the exposure time lower than 0.25 s. However, intentional holding of the blink reflex could lead to potential eye injury. Several measuring instruments and laser pointers are based on Class 2
Class 2M	This class laser is safe as the blink reflex if not viewed through optical instruments. Similar to Class 1M, this class laser lights are with a large divergence, the light passing through the pupil should not exceed the specifications for Class 2
Class 3A	This laser class is safe if handled carefully. The maximum permissible exposure (MPE) can be exceeded, which is associated with a low injury risk

Laser class	Features
Class 3B	This class is hazardous if exposed directly to the eye. Protective eyewear must be used where direct viewing is needed or may occur. The equipment with Class 3B lasers must be fitted with a safety interlock and a key switch
Class 4	This class is the most dangerous among all lasers. This laser can cause permanent eye damage or burn the skin as a result of direct, diffuse or indirect beam viewing or contact. These lasers may cause a fire risk as they can ignite combustible materials. Several laser used in scientific, industrial, military and medical applications fall in this category. The equipment with Class 4 lasers must be fitted with a safety interlock and a key switch

2.3 Classification based on generating media

Based on the generating media, lasers can be classified into three groups: (1) carbon dioxide (CO_2) lasers, (2) neodymium (Nd) lasers and (3) neodymium yttrium-aluminium-garnet (Nd-YAG) lasers (Thawari et al. 2005; Mathew et al.1999; Schuocker 1989). The CO_2 lasers can be used for boring, cutting and engraving (Chow et al. 2012; Kan 2014b). CO_2 lasers can be of the four types: (a) fast axial flow, (b) slow axial flow, (c) transverse flow and (d) slab lasers (Powell1993). In fast axial flow, a mixture of CO_2, nitrogen (N_2) and helium (He) is used at high speed by a turbine. A simple blower is used for slow axial flow and transverse flow lasers. Slab lasers need no pressurization as they work in static condition.

The Nd lasers and Nd-YAG lasers, although similar in style, they are applied in different areas (Schuocker 1989). The former is used for boring, whereas the later is used for engraving. All the laser classes can be used for welding.

Generally, CO_2 lasers are produced by passing the electric discharge through a gaseous mixture excited by direct current (DC) with radio frequency (RF) energy. This method is widely used as the DC-excited designs need electrodes inside the cavity, due to risk of erosion. Industrial cutting of steel, aluminium, paper, plastics, wood, and fabrics can be performed by CO_2 lasers. For cutting metals and ceramics; and scribing Nd-YAG lasers are used.

The laser generator and external optics need cooling while in use (Choudhury and Shirley 2010). This can be achieved by a coolant such as water by circulating through a chiller in the laser equipment. The use of water cools the material and removes the debris. In addition, it helps to achieve parallel kerf and multi-directional cutting at high dicing speeds.

Another type of solid-state laser known as "fibre laser" is now becoming popular in metal cutting industry. This laser is produced from a solid gain medium without the use of any liquid or gas. The major advantage of fibre laser is the extremely small spot size. Fibre lasers, with a wavelength of 1.064 μm, can produce an extremely small spot size, which makes it ideal for cutting reflective metals.

3. APPLICATION OF LASER IN APPAREL INDUSTRY

As an all-new process, there are several applications of laser in apparel industry. Laser engraving and cutting technologies now being widely applied in many garment industries, fabric production units, other textile and leather industries (Choudhury and Shirley 2010; Nayak and Khandual 2010). Various applications of laser are discussed in the following section.

3.1 Fabric fault detection

When fabric is received at the stores of a garment production unit, the faults in the fabric can be detected with morphological image processing based on laser (Mallik-Goswami and Datta 2000; Ribolzi et al. 1993; Mursalin et al. 2005). Laser-based optical Fourier transform analysis can be used for fault detection in the fabric as the pattern is repeated at regular intervals. The fabric is focused with a laser and the diffraction gratings obtained from the periodicity of longitudinal and transverse threads in the fabric are superimposed. A Fourier lens is used to produce the diffraction pattern of the fabric. A second Fourier lens with same focal length magnifies and inverts the test sample image. A charge-coupled device (CCD) camera is used to capture the image. The data is transferred and stored in a computer. The computer programming helps in comparing the acquired images with the stored images by converting the image into binary mode. A fault is reported when the measured parameter is deviating from the standard. The severity of the fault depends upon the amount of deviation from the standard.

3.2 Laser cutting

After they were introduced in the 19th century, the fashion designers are widely adopting laser cutting in garment manufacturing (Petrak and Rogale 2001). In synthetic fabrics, laser cutting produces well-finished edges as the laser melts and fuses the edge, which avoids the problem of fraying produced by conventional knife cutters. Furthermore, use of laser cutting is increasingly used for leather due to the precision of cut components. In fashion accessories such as jewellery, laser cutting can be used to produce new and unusual designs to produce a fusion of apparel design and jewellery style.

In laser cutting a laser is used to cut the fabric into the desired pattern shapes. A very fine laser is focused on to the fabric surface, which increases the temperature substantially and cutting takes place due to vaporization. Normally gas lasers (CO_2) are used for cutting of fabric. The cutting machine (Fig. 2) includes a source of laser, a cutting head fitted with mirrors to reflect the laser beam to the cutting line, a computer to control the entire system and a suitable mean for removing the cut parts. The application of inert gases (N_2, He) during cutting prevents the charring and removes debris and smoke from the cutting area. Like the mechanical cutting devices, a laser beam does not become blunt and need sharpening. Automatic single ply laser cutters are faster (30–40 m/min) than automatic multiple ply knife cutters (5–12 m/min). However, while cutting multiple plies, knife cutters are faster per garment cut and also cheaper.

Figure. 2 A laser-cutting machine

The limitation of laser cutting is the number of lays of the fabric that can be cut by the beam. Best result is obtained while cutting single or a few lays, but the accuracy and precision is not obtained with several plies. In addition there is a chance of the cut edges to be fused together especially in case of synthetics. In some cases the sealing of the edges of cut patterns and sewn garment parts is essential to prevent fraying, where the laser plays the role. As in garment production facilities emphasis is given in multiple lay cutting, the laser cutting seems unlikely to become widespread. However, it is successfully used in cutting of sails where single ply cutting is the norm and a slight fusing of the edge of synthetics and woven materials is desirable. In addition, laser cutting is used in some areas of home furnishing.

Laser cutting is cheaper compared with the traditional cutting methods (Mahrle and Beyer 2009). Furthermore, as the laser cutting doesn't have mechanical action, high precision of the cut components at high cutting speed are feasible (Mathew et al. 1999). The laser cutters are safer and include simple maintenance features, which can be operated for longer duration. The laser cutters can be integrated to the computer technology. It can produce the products at the same time when designing in the computer. Laser cutting machines have faster speed and simpler operation.

Laser cutting machines are suitable for cutting textile fabrics, composites and lather materials (Caprino and Tagliaferri 1988; Steen et al. 2010; Cenna and Mathew 2002). They can operate for a wide range of fabric, which is not possible with die cutters. Hence, laser cutting machines are gradually been accepted in garment manufacturing. The features of laser applications include:

- Laser marking, laser engraving and laser cutting combined in one step
- No mechanical wear, hence good quality
- No fixation of material is required due to force-free processing
- No fabric fraying in synthetic fibres due to formation of fused edges
- It is clean and lint-free
- Simple process due to integrated computer design

- High quality raw materials and significant cost saving
- Extremely high precision in cutting contours
- High working speed
- Contactless, wear-free technique
- No chips, less waste

3.3 Objective evaluation of seam pucker

Garment appearance greatly influences garment quality. Seam pucker negatively affects the garment appearance (Nayak and Padhye 2014b, 2015a; Nayak et al.2010, 2013; Fan and Liu 2000). There are several methods to measure seam pucker, but the conventional rating system developed by American Association of Textile Chemists and Colourists (AATCC) is mainly used. The laser beam can measure the degree of puckering in garments by geometrical models. In this method a seam in the garment is scanned by a 3D laser scanner by putting the garment on a dummy. The laser head can be moved to any 3D space within a confined place by an operator. It is possible to scan the target object from different angles. A pucker profile of the scanned seam can be obtained by processing the image with a 2D digital filter. Physical parameters such as log σ^2 (σ is variance) can be obtained from the pucker profile, which can then be linearly related to grade for seam pucker. From the objectively measured log σ^2, the pucker grade can be objectively evaluated.

3.4 Mass customization

The term mass customization is used when custom-fit garments are obtained depending on the body dimensions and individual's choice. The very first thing to mass customize garments is the accurate measurements of individual's body (Nayak and Padhye et al. 2015a). Laser scanning technology is one of the many techniques used for measurement. Laser scanning technology uses one or multiple thin and sharp stripe lasers to measure body size. Cameras are also used to acquire the scene and assist the laser scanner. The body measurements are derived by applying simple geometrical rules (D'Apuzzo 2007; Tong et al. 2012; Ashdown et al. 2004). In order to confirm the harmlessness of the beam, only eye-safe lasers can be used. Additional optical devices such as mirrors can be used to assist a single laser beam. The laser scanning unit (Fig. 3) consisting of light sensors and optical systems focuses on the human body for digitisation. The number of light sensors and optical systems can vary as per the positions of the body. For example, Vitronic1 body scanner consists of three scanning units that can synchronously move vertically along three pillars.

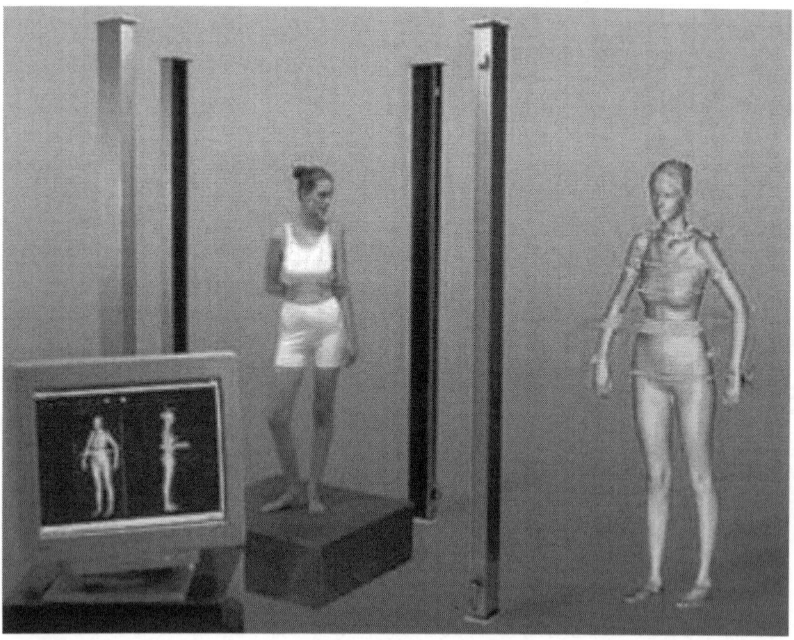

Figure. 3 3D body scanning by Laser scanner

3.5 Laser-based denim fading

Now the age of fading of denim by sandblasting is becoming older as the new technology of laser fading is replacing it (Ortiz-Morales et al. 2003; Tarhan and Sarıışık 2009). In laser fading, a computer drives the laser beam to the material where marking or fading is required. The laser beam decomposes the dye and the resulting vapors are vented away. The material fades only where the beam impacts on the fabric. Commercially two types of lasers are being used: solid based (wavelength of 1 μm) and gas based (wavelength of 10 μm). The desired degree of fading depends upon the wavelength, power density, and pulse width of the laser beam. The method of marking or fading by laser is more environmental friendly as compared to acid washing or sandblasting (Kan et al. 2010). A laser-faded denim sample is shown in the Fig. 4.

Figure. 4 Denim faded by laser

3.6 Laser engraving

In laser engraving laser is used to mark or engrave an object. The process is very complex, and often computerised systems are used to drive the laser head (Kan et al. 2010; Juciene et al. 2013). In spite of the complexity, very precise and clean engravings can be obtained with high rate of production. The technique does not involve physical contact with the engraving surface, hence, no wear and tear. The marks produced by laser engraving are clean, crisp and permanent. In addition, lasers are faster than other conventional methods used for product imprinting, which provides greater versatility in material selection. One machine can be used to cut through thin materials as well as make engravings on them. Laser engraving is used to engrave the printing screens, for hollowing, for creating pattern buttons, to engrave leather, denim etc. (Fig. 5). Pictures, flower patterns and even personalized signatures can be engraved on leather shoes, leather bag, wallet, leather belt, leather sofa and leather clothes, greatly increasing the added value of products. In addition laser engraving is used to create embroidered pattern in the fabric by colour fading and burning the fabrics. The low cost sealed CO_2 lasers are preferred for laser engraving.

Denim engraving is another fast-growing application of laser using sealed CO_2lasers (Juciene et al. 2013; Kan 2014a). The laser is used to create minute designs and patterns on denim fabric as well as finished denims. This technic can be used in place of the traditional techniques such as sandblasting and acid washing. The accuracy and design flexibility is very wide, which can't be achieved by the traditional methods. Lasers can produce 3D effects by techniques such as embroidering, embossing, or even apparent cuts, tears and mends. Any image that is created in a computer aided design (CAD), can be transferred to denim by suitable laser process. While using lasers, features such as good mode quality, high power stability, real-time control of laser power and fast pulse rise-time are the important parameters that can lead to colour change without charring or other damage to the fabric. Such damage could reduce the product life and cosmetically unacceptable. The advantages of laser engraving over traditional methods include:

- High working speed without mechanical contact
- No wear and tear of components
- Reduced waste
- Complete exhaust and filtering
- Exact contours possible

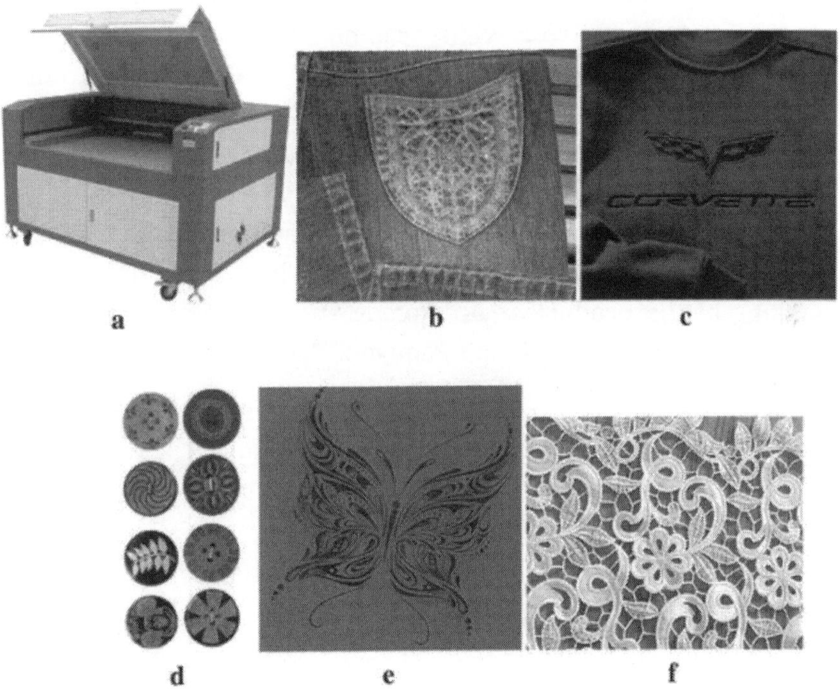

Figure. 5 Laser engraving items: **a** engraving machine, **b** denim, **c** garment, **d** buttons, **e**leather and **f** embroidery

3.7 Welded garment production

Welding is an alternative process of joining fabrics for garment production where the thermoplastic materials are joined together by the application of heat. The heat can be supplied by ultrasonic or by high powerful laser (Petrie 2015). The welded garment though weaker than the sewn counterpart, gives better appearance as it does not contain bulky seam and is more flexible.

3.8 Bar code scanning

The scanners used to scan the barcodes for product identification typically uses helium–neon (He–Ne) lasers. The laser beam bounces out of a rotating mirror while scanning the code. This sends a modulated beam to a computer, which contains the product information. Semiconductor based-lasers can also be used for this purpose. However, some of the recent manufacturers are using Radio Frequency Identification (RFID) based tags instead of barcodes due to certain advantages (Nayak and Padhye 2014c). The RFID tag can be processed quickly and it avoids the physical handling of the product as in barcode systems (Nayak et al.2007, Nayak et al. 2015b; Nayak and Padhye 2015b). The mechanism of a bar code scanner is shown in Fig. 6.

3.9 Laser marking

Laser can also be used in marking on various surfaces. The advantages of laser marking include fast, high precision and clear marking on products of varying contour and hardness. It can also be used for a wide range of organic polymers where precession can be obtained even with complex designs. Laser marking is durable and can be applied in clothing, leather and metals. Laser marking is considered to be the best choice for branded clothing and marking fashion accessories during processing.

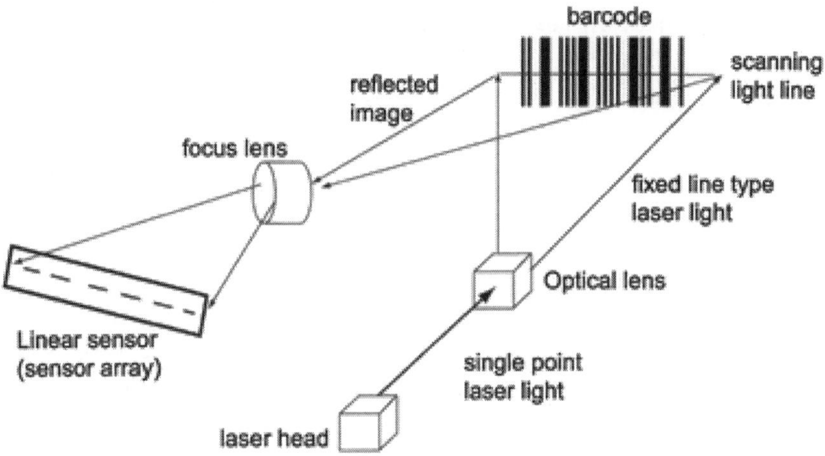

Figure. 6 A laser bar code scanner

3.10 Miscellaneous uses

There are many other applications of laser in apparel industry as discussed below:

- In some stitching machines (e.g., automatic welt pocket attaching machine) laser beam is engaged for automatic and accurate positioning of the welt and to make a slit across the fabric.
- In some metal detection machines laser beam is employed to find if any needle part is there in the final finished garment.
- Laser-engraving machine can engrave any decoration on the surface layer of any material, making the products looking high grade and exquisite.
- Laser technology is gaining impetus in garment finishing which can produce various surface ornamentations without any wet processing. This technique is very accurate and can work fast with good repeatability and reproducibility (Hung et al.2011; Kan et al. 2010).

- The application of antimicrobial finishes into textiles has significantly improved compared to the past (Nayak and Padhye 2014a; Nayak et al. 2008). Laser treatment was used for durable antibacterial properties on cotton fabric using silver nanoparticles by Nourbakhsh et al. (Nourbakhsh and Ashjaran 2012).

4. HAZARDS BY LASER AND THEIR CONTROL

4.1 Electrical shock

More people are being killed by electrocution from the laser electronics than blinded from exposure to a laser beam (Sliney 1995). Lethal voltages are present in the power supplies of lasers. If one is not experienced working with high voltages in general and laser power supplies in particular, then the person should not be allowed to carry out any work with the laser. During maintenance, the power supply should be unplugged from its electrical outlet.

4.2 Eye injury potential

Eye is the most vulnerable body part to a laser beam (Henderson et al. 2003; Buckley 2010; Barkana and Belkin 2000). With certain lasers severe damage can be caused due to the high concentration of laser energy on retina (Wyrsch et al. 2010). Class 4 lasers can damage the tissues in the eye interior. The class and duration of laser exposure are the deciding factors in the eye injury. For example, no injury is be expected while working with the laser with wavelength in the visible spectrum (400–700 nm). Lasers of Class 3B or Class 4 can lead to an eye injury before the aversion response can protect the eye. Tables 2 and 3 describe various types of laser hazards and precautions to be taken as specified by American National Standards

Institute (ANSI) and Occupational Safety and Health Administration (OSHA).

Table 2 Nature of hazards and precautions for various classes of lasers

Laser class	Nature of hazard (ANSI and OSHA)	Precautions
Class 1	Low power, safe to view	No potential hazard
Class 1M	Low power, hazardous when viewed directly for longer than 1000 s	Do not view directly with optical instruments
Class 2	Low power, hazardous when viewed directly for longer than 0.25 s	Do not stare into the beam
Class 2M	Medium power, nonhazardous when viewed directly for less than 0.25 s	Do not stare into the beam, do not view directly with optical instruments
Class 3A	Medium power (0.5 W), are hazardous when viewed directly	Avoid direct eye exposure
Class	Medium power (0.5 W), are hazardous when viewed	Avoid direct eye exposure

Laser class	Nature of hazard (ANSI and OSHA)	Precautions
3B	directly	
Class 4	High power (>0.5 W), produce ocular, skin and fire hazards	Avoid eye or skin exposure to direct or scattered radiation

Table 3 Harmful effect of laser radiation on eye and skin

Laser type	Wavelengths (nm)	Impacts on eye	Impacts on skin
Excimer laser	100–315	Cornea ignition	Sunburn, accelerated aging
He–Ne laser	315–380	Lens opacity	Increased pigmentation
Nd-YAG laser	380–780	Violation of the retina	Darkening of pigment, burns
High performance diode laser	780–1400	Lens opacity, violation of the retina	Burns
CO_2 laser	1400–3000	Lens opacity, burning of	Burns

Laser type	Wavelengths (nm)	Impacts on eye	Impacts on skin
		the cornea	
CO_2 laser	3000–10,000	Burning of the cornea	Burns

Various standards are devised around the world for the laser related hazards. The standards devised by ANSI are shown in Table 4.

Table 4 Various ANSI standards dealing with laser hazards

Laser standards	Description
ANSI Z136.1 (safe use of lasers)	This standard is the foundation of laser safety programs for industry, military, research and development and higher education (universities)
ANSI Z136.2 (safe use of optical fibre communication systems utilizing laser diode and LED sources)	This standard provides guidelines for the safe use, maintenance, service and installation of optical systems utilizing laser diodes or light emitting diodes operating at wavelengths between 0.6 µm and 1 mm
ANSI Z136.3 (safe use of lasers in health	This standard provides guidelines for individuals working with Class 3B and

Laser standards	Description
care)	Class 4 lasers and laser systems in health care
ANSI Z136.4 (recommended practice for laser safety measurements for hazard evaluation)	This standard provides guidelines for measurement procedures necessary for the classification and evaluation of optical radiation hazards
ANSI Z136.5 (Safe use of lasers in educational institutions)	This standard provides guideline for organisations and implementation of laser safety and training programs. In addition, it privides graphics for entryway controls, laser installations and laser laboratory layouts
ANSI Z136.6 (safe use of lasers outdoors)	This standard provides guidelines for the safe use of lasers in an outdoor environment such as construction, light shows, scientific, research and military
ANSI Z136.7 (testing and labelling of laser protective equipment)	This standard provides guidelines on the test methods and protocols used to provide eye protection from lasers and laser systems
ANSI Z136.8 (safe use of lasers in research, development, or	This standard provides guidelines on the safe use of lasers and laser systems found in research, development or testing environments, where safety

Laser standards	Description
testing)	controls common for commercial lasers may either be missing or disabled
ANSI Z136.9 (safe use of lasers in manufacturing environments)	This standard provides guidelines for laser exposures when lasers are used in manufacturing environments. This also includes policies and procedures for safety in both public and private industries as well as product development along with testing

4.3 Skin injury potential

The injury to the skin due to laser radiation is less severe compared to the eye. However, the chances of exposure of skin is higher than that of the eye due to its greater surface area (Yashima et al. 1991; Sliney 1995). The eye injury is more significant than the skin as the loss in the vision is irreparable. In normal laser working condition there is very less chance that a large area of the skin is exposed. The injury to the skin due to laser exposure can be divided into two categories: (a) thermal injury and (b) photochemical induced injury. The former is caused by the high power laser beams, whereas the latter is caused by exposure to scattered ultraviolet (UV) laser radiation (Bernstein et al. 1997). Direct contact or exposure to the laser beam can lead to thermal injuries. Although these injuries are painful, usually they are not serious and proper beam management and hazard awareness can prevent these. In addition, specular or even diffused reflections can lead to photochemical injury over time.

The objective of hazard control methods is primarily stop the laser contacting the skin or entering into the eye (Protection 1996). These control methods can be grouped into three sections such as: (a) administrative controls [labels, signs, standard operating procedures (SOPs), etc.], (b) engineering controls (barriers, blocks, etc.) and (c) protective controls (eyewear, uniform, etc.).

4.4 Administrative controls

The management should only allow the trained persons to work on laser equipment. The operator should follow the instructions as in SOPs. The laser equipment should be switched off while not being used. All laser equipment of Class 3B or Class 4 lasers need to be labelled with "Danger" symbol, specifying the laser class. A Class 3B laser device should mention "LASER RADIATION-AVOID DIRECT EXPOSURE TO THE BEAM", which must be written above the tail of the sunburst. A Class 4 laser device should mention "LASER RADIATION-AVOID EYE OR SKIN EXPOSURE TO DIRECT OR SCATTERED RADIATION" above the tail of the sunburst.

4.5 Engineering controls

The possibility of accidental exposures to laser hazards can be best controlled by engineering controls. All the Class 3B and Class 4 laser equipment should prevent the access of unauthorized personnel in the working area while laser is operational. All the Class 3B and Class 4 lasers should have a non-flammable cover sufficient to hold the excitation device and the beam. The laser systems can be fitted with key switches or password protected for better safety. In the laser chamber, the setting should guarantee no direct eye contact. The most hazardous aspect of laser use is the beam alignment, where most eye injuries occur. Hence to avoid this, the instructions described in the SOP must be understood. The lowest visible beam power should be used for beam alignment.

4.6 Protective equipment controls

The user of laser equipment should wear appropriate personal protective clothing/device for eye and skin protection during initial setting as well as the normal working (Nayak et al. 2015c). The skin covering and the eyewear protect the skin and eye, respectively from direct exposure.

5. CONCLUSIONS AND FUTURE DIRECTIONS

Laser technology can be used for various applications on materials ranging from metals to textiles with noncontact patterns. In garment production, it can be applied onto different products ranging from home textiles to fashion accessories. In garment manufacturing, CO_2 gas lasers have wide and successful applications. Laser technique, is entirely different from traditional textile processes, as it has the flexibility in design and operation without any pollution or waste material. There are several other advantages of using laser over the conventional methods in cutting, engraving, embossing, denim fading etc. In addition, laser involves lower risk of product damage, use of low consumables and free from disposing of toxic by-products, as there may be with some methods. The laser equipment of today has gradually evolved from those used in early days. The old laser equipment were difficult to run, cumbersome and hard to maintain. However, the modern equipment are easy to operate, simple to learn and easy to maintain. Furthermore, the earlier equipment had several safety issues and they needed replenishment of gasses at regular intervals. The garment production units should take the advantage of applying laser in the post multi-fibre agreement regime to produce more competitive products.

AUTHORS' CONTRIBUTIONS

RN designed the paper, performed the literature survey and completed the paper. RP guided during the paper design and helped in feedback and comments in final format of the paper.

COMPETING INTERESTS

The authors declare that they have no competing interests.

REFERENCES

1. Ashdown, S. P., Loker, S., Schoenfelder, K., & Lyman-Clarke, L. (2004). Using 3D scans for fit analysis. Journal of Textile and Apparel, Technology and Management, 4, 1–12.

2. Bahtiyari, M. (2011). Laser modification of polyamide fabrics. Optics & Laser Technology, 43, 114–118.

3. Barkana, Y., & Belkin, M. (2000). Laser eye injuries. Survey of Ophthalmology,44, 459–478.

4. Belli, R., Miotello, A., Mosaner, P., & Toniutti, L. (2005). Laser cleaning of ancient textiles. Applied Surface Science, 247, 369–372.

5. Bernstein, L. J., Kauvar, A. N., Grossman, M. C., & Geronemus, R. G. (1997). The short-and long-term side effects of carbon dioxide laser resurfacing. Dermatologic Surgery, 23, 519–525.

6. Buckley, E. (2010). Eye-safety analysis of current laser-based scanned-beam projection systems. Journal of the Society for Information Display, 18, 944–951.

7. Caprino, G., & Tagliaferri, V. (1988). Maximum cutting speed in laser cutting of fiber reinforced plastics. International Journal of Machine Tools and Manufacture, 28, 389–398.

8. Cenna, A., & Mathew, P. (2002). Analysis and prediction of laser cutting parameters of fibre reinforced plastics (FRP) composite materials. International Journal of Machine Tools and Manufacture, 42, 105–113.

9. Choudhury, I., & Shirley, S. (2010). Laser cutting of polymeric materials: an experimental investigation. Optics & Laser Technology, 42, 503–508.

10. Chow, Y. F., Chan, A., & Kan, C.-W. (2012). Effect of CO2 laser irradiation on the properties of cotton fabric. Textile Research Journal, 82, 1220–1234.

11. D'Apuzzo, N. (2007). 3D body scanning technology for fashion and apparel industry. In Electronic Imaging 2007 (vol. 6491, pp. 649100–649102). International Society for Optics and Photonics.

12. Dowden, J. (2009). The Theory of Laser Materials Processing: Heat and Mass Transfer in Modern Technology. Berlin: Springer Science & Business Media.

13. Fan, J., & Liu, F. (2000). Objective evaluation of garment seams using 3D laser scanning technology. Textile Research Journal, 70, 1025–1030.

14. Henderson, R., & Schulmeister, K. (2003). Laser safety. Boca Raton: CRC Press.

15. Hung, O., Song, L., Chan, C., Kan, C., & Yuen, C. (2011). Using artificial neural network to predict colour properties of laser-treated 100% cotton fabric. Fibers and Polymers, 12, 1069–1076.

16. Istook, C. L., & Hwang, S.-J. (2001). 3D body scanning systems with application to the apparel industry. Journal of Fashion Marketing and Management, 5, 120–132.

17. Juciene, M., Urbelis, V., Juchnevičienė, Ž., & Čepukonė, L. (2013). The effect of laser technological parameters on the color and structure of denim fabric. Textile Research Journal. doi:10. 1177/0040517513494256.

18. Kan, C.-W. (2014a). CO2 laser treatment as a clean process for treating denim fabric. Journal of Cleaner Production, 66, 624–631.

19. Kan, C. (2014b). Colour fading effect of indigo-dyed cotton denim fabric by CO2laser. Fibers and Polymers, 15, 426–429.

20. Kan, C. (2015). Washing techniques for denim jeans (p. 313). Denim: Manufacture, Finishing and Applications.

21. Kan, C., Yuen, C., & Cheng, C. (2010). Technical study of the effect of CO2 laser surface engraving on the colour properties of denim fabric. Coloration Technology, 126, 365–371.

22. Kovacs, L., Zimmermann, A., Brockmann, G., Gühring, M., Baurecht, H., Papadopulos, N., et al. (2006). Three-dimensional recording of the human face with a 3D laser scanner. Journal of Plastic, Reconstructive and Aesthetic Surgery,59, 1193–1202.

23. Lu, J.-M., Wang, M.-J. J., Chen, C.-W., & Wu, J.-H. (2010). The development of an intelligent system for customized clothing making. Expert Systems with Applications, 37, 799–803.

24. Lucas, J., Belino, N., Miguel, R., Pereira, M., & Ribeiro, L. (2015). Digital printing techniques for denim jeans (p. 287). Denim: Manufacture, Finishing and Applications.

25. Mahrle, A., & Beyer, E. (2009). Theoretical aspects of fibre laser cutting. Journal of Physics. D. Applied Physics, 42, 175507.

26. Mallik-Goswami, B., & Datta, A. K. (2000). Detecting defects in fabric with laser-based morphological image processing. Textile Research Journal, 70, 758–762.

27. Mathew, J., Goswami, G., Ramakrishnan, N., & Naik, N. (1999). Parametric studies on pulsed Nd: YAG laser cutting of carbon fibre reinforced plastic composites. Journal of Materials Processing Technology, 89, 198–203.

28. Mursalin, T. E., Eishita, F. Z., & Islam, R. (2005). Fabric defect inspection system using neural network and microcontroller. Statistics, 50, 100.

29. Nayak, R., Chatterjee, K., Khandual, A., & Jajpura, L. (2008). Evaluation of functional finishes-An overview. Man-Made Textiles in India, 51, 131–135.

30. Nayak, R., Chatterjee, K., Khurana, G., & Khandual, A. (2007). RFID: tagging the new era. Man-Made Textiles in India, 50, 174–177.

31. Nayak, R., & Khandual, A. (2010). Application of laser in apparel industry.Colourage, 57, 85–90.

32. Nayak, R., & Padhye, R. (2014a). Antimicrobial finishes for textiles. In R. Paul (Ed.), Functional Finishes for Textiles: Improving Comfort, Performance and Protection. Amsterdam: Elsevier.

33. Nayak, R., & Padhye, R. (2014b). The care of apparel products. In R. Sinclair (Ed.),Textiles and fashion: Materials, design and technology. United Kingdom: Elsevier.

34. Nayak, R., & Padhye, R. (2014c). Introduction: the apparel industry. In R. Nayak & R. Padhye (Eds.), Garment Manufacturing Technology. Amsterdam: Elsevier.

35. Nayak, R. & Padhye, R. (2015a). Care labeling of clothing. In R. Nayak. & R. Padhye (Eds.), Garment Manufacturing Technology. Amsterdam: Elsevier

36. Nayak, R., & Padhye, R. (2015b). Garment Manufacturing Technology. Amsterdam: Elsevier.

37. Nayak, R., Padhye, R., Dhamija, S., & Kumar, V. (2013). Sewability of air-jet textured sewing threads in denim. Journal of Textile and Apparel, Technology and Management, 8, 1–11.

38. Nayak, R., Padhye, R., & Gon, D. P. (2010). Sewing performance of stretch denim.Journal of Textile and Apparel, Technology and Management, 6(3), 1–9.

39. Nayak, R., Padhye, R., Wang, L., Chatterjee, K., Gupta, S. (2015a). The role of mass customisation in the apparel industry. International Journal of Fashion Design, Technology and Education, 8, 162–172.

40. Nayak, R., Singh, A., Padhye, R., & Wang, L. (2015b). RFID in textile and clothing manufacturing: technology and challenges. Fashion and Textiles, 2, 1–16.

41. Nayak, R., Padhye, R., & Wang, L. (2015c). How to dress at work. In S. Patole (Ed.), Management and Leadership–A Guide for Clinical Professionals, Germany: Springer

42. Nourbakhsh, S., & Ashjaran, A. (2012). Laser treatment of cotton fabric for durable antibacterial properties of silver nanoparticles. Materials, 5, 1247–1257.

43. Ondogan, Z., Pamuk, O., Ondogan, E. N., & Ozguney, A. (2005). Improving the appearance of all textile products from clothing to home textile using laser technology. Optics and Laser Technology, 37, 631–637.

44. Ortiz-Morales, M. N., Poterasu, M., Acosta-Ortiz, S. E., Compean, I., & Hernandez-Alvarado, M. R. (2003). A comparison between characteristics of various laser-based denim fading processes. Optics and Lasers in Engineering, 39, 15–24.

45. Ozguney, A. T. (2007). The comparison of laser surface designing and pigment printing methods for the product quality. Optics and Laser Technology, 39, 1054–1058.

46. Petrak, S., & Rogale, D. (2001). Methods of automatic computerised cutting pattern construction. International Journal of Clothing Science and Technology,13, 228–239.

47. Petrie, E. (2015). Alternative fabric-joining technologies 13. In R. Nayak, R. Padhye (Eds.), Garment Manufacturing Technology. Cambridge, UK: Elsevier.

48. Potluri, P., & Atkinson, J. (2003). Automated manufacture of composites: handling, measurement of properties and lay-up simulations. Composites Part A Applied Science and Manufacturing, 34, 493–501.

49. Powell, J. (1993). CO2 laser cutting. Cambridge: Univ Press.

50. Protection, I. C. O. N.-I. R. (1996). Guidelines on limits of exposure to laser radiation of wavelengths between 180 nm and 1,000 [mu] m. Health Physics, 71, 804–819.

51. Ready, J. F., Farson, D. F., & Feeley, T. (2001). LIA handbook of laser materials processing. Orlando: Laser Institute of America.

52. Ribolzi, S., Merckle, J., Gresser, J., & Exbrayat, P. (1993). Real-time fault detection on textiles using opto-electronic processing. Textile Research Journal, 63, 61–71.

53. Schuocker, D. (1989). Laser cutting. Material and Manufacturing Process, 4, 311–330.

54. Simmons, K. P., & Istook, C. L. (2003). Body measurement techniques: comparing 3D body-scanning and anthropometric methods for apparel applications. Journal of Fashion Marketing and Management, 7, 306–332.

55. Sliney, D. H. (1995). Laser safety. Lasers in Surgery and Medicine, 16, 215–225.

56. Steen, W., Watkins, K. G., & Mazumder, J. (2010). Laser material processing. Amsterdam: Springer Science & Business Media.

57. Sutcliffe, H., Cooper, M., & Farnsworth, J. (2000). An initial investigation into the cleaning of new and naturally aged cotton textiles using laser radiation. Journal of Cultural Heritage, 1, S241–S246.

58. Tarhan, M., & Sarıışık, M., (2009). A comparison among performance characteristics of various denim fading processes. Textile Research Journal, 79, 301–309.

59. Thawari, G., Sundar, J. S., Sundararajan, G., & Joshi, S. (2005). Influence of process parameters during pulsed Nd: yAG laser cutting of nickel-base superalloys.Journal of Materials Processing Technology, 170, 229–239.

60. Tong, J., Zhou, J., Liu, L., Pan, Z., & Yan, H. (2012). Scanning 3d full human bodies using kinects. Visualization and Computer Graphics, IEEE Transactions on, 18, 643–650.

61. Vilumsone-nemes, I. (2012). Industrial Cutting of Textile Materials. Amsterdam: Elsevier.

62. Wyrsch, S., Baenninger, P. B., & Schmid, M. K. (2010). Retinal injuries from a handheld laser pointer. New England Journal of Medicine, 363, 1089–1091.

63. Yashima, Y., McAuliffe, D. J., Jacques, S. L., & Flotte, T. J. (1991). Laser-induced photoacoustic injury of skin: effect of inertial confinement. Lasers in Surgery and Medicine, 11, 62–68.

64. Yuan, G. X., Jiang, S. X., Newton, E., Fan, J. T., & Au, W. M. (2012). Application of laser treatment for fashion design. Journal of the Textile Institute, 103, 48–54.

10

DIODE LASER ASSISTED FILAMENT WINDING OF THERMOPLASTIC MATRIX COMPOSITES

Fabrizio Quadrini *, Erica Anna Squeo and Claudia Prosperi

Department of Mechanical Engineering, University of Rome Tor Vergata, Via del Politecnico 1, 00133 Rome, Italy

ABSTRACT

A new consolidation method for the laser-assisted filament winding of thermoplastic prepregs is discussed: for the first time a diode laser is used, as well as long glass fiber reinforced polypropylene prepregs. A consolidation apparatus was built by means of a CNC motion table, a stepper motor and a simple tensioner. Preliminary tests were performed in a hoop winding configuration: only the winding speed was changed, and all the other process parameters (laser power, distance from the laser focus, consolidation force) were kept constant. Small wound rings with an internal diameter of 25 mm were produced and compression tests were carried out to evaluate the composite agglomeration in dependence of the winding speed. At lower winding speeds, a strong interpenetration of adjacent layers was observed.

KEYWORDS

diode laser; laser processing; thermoplastic prepreg; filament winding; composite material; thermoplastic matrix composites

1. INTRODUCTION

Thermoplastic matrix composites are substituting traditional long glass fiber composites made of thermosetting matrix for the production of small components in aerospace applications. The main advantages associated with thermoplastic matrix composites are the high shelf life of the prepregs, the higher reliability of the production processes and the simplification of the molding operations. However, typical thermosetting matrix composites processes have to be re-designed for the application to this new class of materials. An important example deals with filament winding process. Regarding thermosetting matrix composites a continuous prepreg is wound on a machine mandrel and subsequently cured in an oven. The typical material adhesivity (known as "tack") plays an important role in the material consolidation before curing in the oven. Thermoplastic prepregs have no tack properties and have to be joined during the winding operation. The authors have studied a laser-assisted process to allow the thermoplastic prepreg to adhere to the underlying ply simultaneously with its deposition. Use of a laser source to consolidate thermoplastic prepregs is not a novel technology. In 1992, Agarwal *et al.* already proposed the thermal characterization of the laser-assisted consolidation process [1]. They discussed that this kind of technology was able to produce thermoplastic composite parts of excellent quality and structural integrity, combining the advantages of the traditional thermosetting filament winding technology with the faster processing times of thermoplastic prepregs. The experimental apparatus for the laser-assisted consolidation process consisted of a mandrel mounted on a rotational table, a fixed compaction roller, and an 80 W CO_2 laser. Two ZnSe lenses were used to focus an 8 mm diameter beam into a 6

× 0.5 mm^2 planar beam, and no winding tension was applied. Experimental tests were carried out on continuous carbon fiber reinforced PEEK prepregs. In 1997, Rosselli *et al.* used a similar consolidation apparatus to process the same material, deepening the effect of the process parameters on the properties of the consolidated composite [2]. They chose to apply a high compaction force in combination with the pre-heating of the compaction-roller: as a result, consolidated composites showed a typical shear strength about 50% of the autoclave processed composites. In 1999, the same consolidation apparatus was used again to process wider prepreg tapes made of carbon reinforced PPS [3]. After that, no further studies about laser-assisted filament winding of thermoplastic prepregs were carried out, even though new interesting materials and laser sources entered the market. In 2003, Wang and Lou proposed a thermal model for on-line laser curing of thermoset composites in filament winding, but no experimental tests were performed [4]. No mention is given to laser-processing of long glass fiber reinforced polypropylene (PP) prepregs, although this new composite material has gained high interest in the scientific world. In 2000, Torres and Bush studied the sheet extrusion and thermoforming of discrete long glass fiber reinforced PP [5]. During extrusion, the long glass fibers were organized into coherent fiber mats which persisted into the solid state, and were able to withstand the deformation process by thermoforming. In 2004, Zampaloni *et al.* presented the stamp thermoforming process which involved supporting the thermoplastic sheet with a bed of heated viscous fluid that applied a hydrostatic pressure across the part throughout the forming process [6]. Experiments were performed on a random oriented continuous glass-mat fiber reinforced PP by means of a hemispherical punch. Dealing with filament winding, numerical simulations are mainly discussed: a first contribution was given by Schlottermüller *et al.* in their study about the simulation of thermal residual stresses in thermoplastic filament winding [7]. Experimental tests were carried out on carbon fiber reinforced PEEK and glass fiber reinforced PP; a gas torch was used as thermal source for the prepreg consolidation in combination with a consolidation roller. Subsequently (2005), this study was extended to evaluate the effect of the tape tension on residual stress and a tape tensioner was added to the consolidation apparatus [8]. Similar experiences were also made by Toso *et al.* who discussed computational simulations and experimental validations for the winding process of glass fiber reinforced PP tapes [9]; also in this case a hot gas torch was used as thermal source. More recently, Antunes *et al.* proposed finite element modeling of thermoplastic matrix composite gas cylinders [10]. Continuous glass/PP commingled fiber tapes were used to produce wrapped pressure gas vessels for domestic applications.

In this study, the authors discuss for the first time the laser-assisted filament winding of continuous glass fiber reinforced PP tapes. A diode laser source was chosen for the composite consolidation, and the easiest winding configuration (*i.e.*, hoop winding) was chosen for preliminary tests. Diode lasers have some industrial advantages compared with other laser systems, namely lower cost and higher efficiency, which permits lower electrical energy consumption. Thanks to their reduced size, an easier integration in production lines is possible. Diode lasers require lower maintenance, guaranteeing a minimum working life of 12,000 hours. Principal disadvantages are related to the low specific power and high spot dimension. Cutting and high speed deep penetration welding are not possible for common metals, but a series of interesting applications are possible for polymer forming and processing. In fact, polymer materials need time to be processed and require low energy; moreover the large spot size is useful to enlarge the laser-processed zone. Experimental tests were performed on small composite rings produced by hoop winding by changing the winding speed but keeping constant the laser power and the focalization condition. Mechanical tests were performed to evaluate the performances of the final filament wound tube portions.

2. EXPERIMENTAL METHODS

The experimental apparatus for the winding process is shown in Figure 1. A diode laser source was fixed on an external frame as well as a CNC motion table for the sample movement along the x and z direction. A stepper motor was mounted on the motion system so as to provide a rotational speed to a steel tube with the external diameter of 20 mm. The adopted laser was a diode laser (Rofin-Sinar, DL 015) with a 940 nm wavelength and a rectangular spot (3.8 × 1.2 mm^2) due to the superposition of two different rays, each one coming from a 750 W emitter diode. A 63 mm long focus lens was used to maximize the depth of field. A flux of nitrogen gas was used to protect the lens during working.

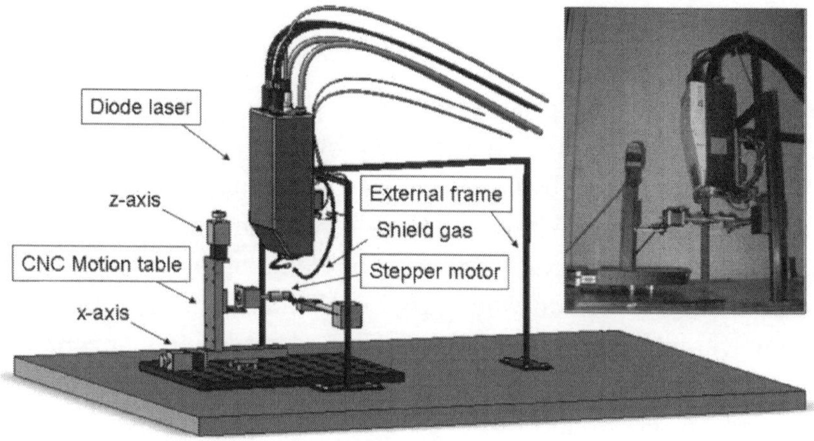

Figure 1. Experimental apparatus for laser-assisted winding of thermoplastic prepregs.

Figure 2 shows a detail of the consolidation apparatus regarding the processing area. A narrow prepreg tape was wound around the steel tube (*i.e.*, the mandrel) and was forced to pass through two polyethylene sheets mechanically clamped together. Due to the friction of the two polyethylene sheets, a consolidation force was applied to the prepreg during winding.

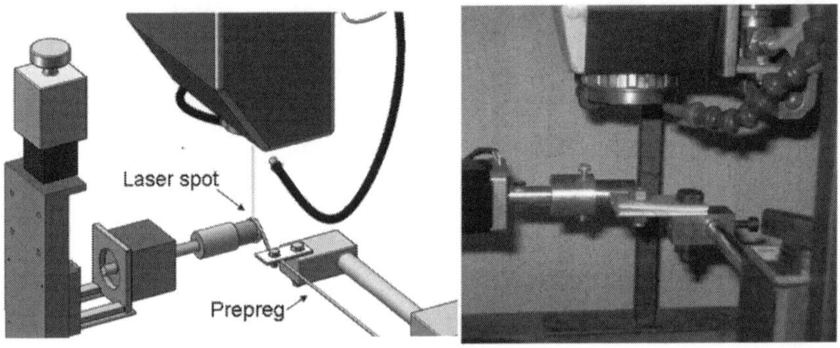

Figure 2. A detail of the consolidation apparatus.

In the experimental procedure, the prepreg tape was wound for a round on the mandrel and shaped by means of a heater. That operation was made by hand, and was necessary for the subsequent positioning under the laser source otherwise the tape stiffness made the laser focalization impossible. After clamping the mandrel to the stepper motor, the non-wound part of the tape was placed between the polyethylene sheets that were clamped in turn. In that condition, the laser-assisted winding process could start: the laser spot was focused on the wound prepreg by means of the z-axis of the CNC table. The laser power was fixed at 50 W and 3 complete revolutions were made without changing the distance from the laser focus. After that, the laser beam was focused again on the external surface of the wound prepreg and other 3 revolutions were performed with the same laser power and rotational speed. Different winding tests were made by only changing the winding speed (14.4, 16.2 and 18 deg/s) because of the very high sensitivity to the laser power: three samples were produced for each winding speed. Higher power values easily led to material degradation or excessive melting, whereas lower power values resulted in poor consolidation.

In order to evaluate the consolidation force during winding, a simple experimental test was performed. A prepreg tape was clamped to a universal testing machine (Alliance RT/50 by MTS) as well as the tensioner. A constant rate was applied to the prepreg and the friction force was measured. The test rate was fixed equal to the peripheral velocity of the tape during the winding tests, and a small change of the friction force in dependence of the speed rate was observed in the experimented range. For example, Figure 3 shows the friction force at the test rate of 210 mm/min which corresponds to the winding speed of 16.2 deg/s. As expected, a maximum is observed (about 3 N) at low displacements, and a plateau value (about 0.5 N) at higher displacements. On average, a consolidation force about 1 N was observed during the 160 mm long test.

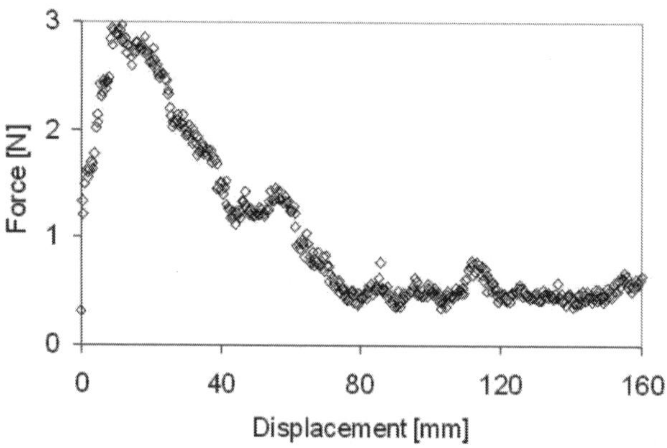

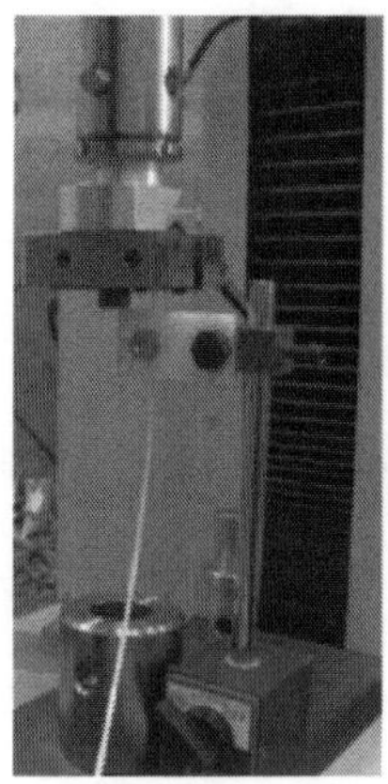

Figure 3. Evaluation of the consolidation force.

Unidirectionally reinforced thermoplastic prepregs (by Jonam Composites Ltd, UK) were used to produce the composites. The GPP45 glass/polypropylene prepreg was supplied in the shape of a long tape, 95 mm in width and 0.35 mm in thickness, with the nominal glass content of 45 vol %. Some narrow tapes, 4 mm wide, were cut from the 95 mm tape and used for winding. As the laser spot was focused on the prepreg surface during laser processing, the maximum axis of the spot was aligned with the tape width, so as to cover the entire tape width by the laser spot. This way, a single laser pass was able to consolidate the entire processed tape. The provided material was qualified by differential scanning calorimetry (DSC): DSC scans (by DSC7, Perkin Elmer) were carried out from room temperature to 170 °C at 10 °C/min. Figure 4 shows the melting peak of the PP matrix for the provided prepreg. As the matrix melting is necessary to obtain a complete composite consolidation, a temperature near the peak temperature has to be reached during laser heating. High temperatures would result in an excess of the material flow during forming and also in matrix degradation. On the contrary, low temperatures would be not sufficient for the composite consolidation.

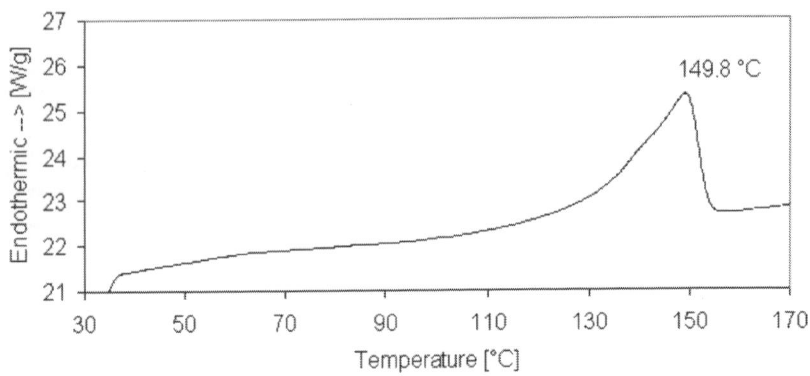

Figure 4. DSC scan of the thermoplastic prepreg.

Mechanical performances of the wound composites were evaluated by means of compression tests which were performed orthogonally to the cylinder axis at the test rate of 10 mm/min up to the maximum displacement of 10 mm.

3. EXPERIMENTAL RESULTS

Depending on the winding speed, different results were obtained in terms of composite aesthetics, thickness and mechanical performances. Figure 5a shows the appearance of a typical consolidated composite (at the winding speed of 18 deg/s). All the consolidated composites showed a good adhesion of the prepreg tape along its revolutions. Evidently, the adopted process conditions (*i.e.*, the low laser power, the focalization condition) were sufficient to cause the polypropylene to melt, and the applied force was sufficient to produce the matrix flow necessary for the consolidation. In particular, a very low consolidation force (on average 1 N) was used in combination with a very low laser power (50 W). Low laser power values are generally applied when a good radiation absorption is exhibited by the processed materials. In this case, not only the prepreg but also the underlying steel tube cooperate for the success of the laser-assisted forming operation. The efficiency of the consolidation phase is clearly shown by the mechanical tests. Figure 5b shows a composite sample during compression: A high deformation is applied to the sample without observing any delamination or break.

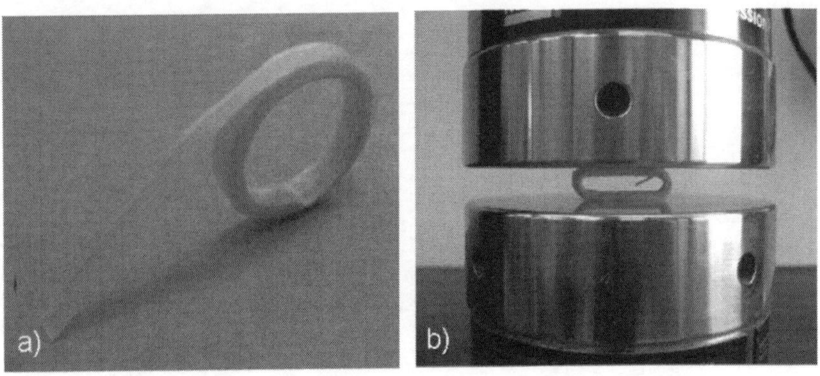

Figure 5. (a) Wound composite at the end of the consolidation process and (b) during compression.

Quantitative results are shown in Figure 6a and 6b in terms of compression curves. Figure 6a shows the dependence of the mechanical performance of the composite samples on the winding speed. In a previous study, the authors already discussed the results of compression tests on thin composite tubes which were produced by molding the same PP prepregs used in the current study [11]. Even if longer tubes were formed, a similar mechanical behavior was observed. By increasing the winding speed, the compression curve decreases. In all the cases, sample breakage was never observed because of the ductile behavior of the thermoplastic matrix composite. However, more reliable results were observed at the highest value of the winding speed: Figure 6b shows the comparison between two composite samples processed at 18 deg/s. Even if the overall reliability has to be improved, at lower winding speed much lower reliabilities were observed. The material degradation was always avoided, but at lower winding speed the input energy increased and the excess in the material flow made the control of the consolidation process impossible. At high energy inputs, the polymer viscosity is so low that the friction force of the tensioner generates a strong interpenetration of the adjacent layers. Probably the best way to consolidate the prepreg tape is to form the material in a semisolid state: the PP matrix has to be soft enough to make the layer joining possible, but not so much to squeeze away from the consolidated layers.

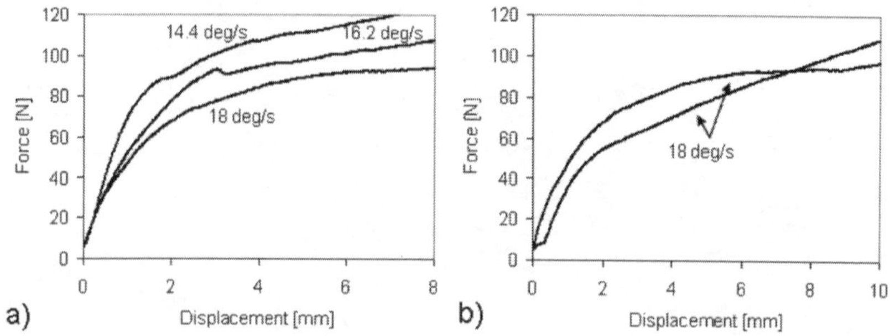

Figure 6. (a) Effect of the winding speed on the compression test and (b) reliability of the results at the highest winding speed.

This statement is confirmed by Figure 7a and 7b, where the final thickness of the composite samples and the appearance of a composite sample processed at 16.2 deg/s are shown, respectively. Unfortunately, a high thickness variation was observed in the composite rings: a similar variation was also observed along the circumference of each single ring. By increasing the winding speed, the composite thickness increases because of the lower interpenetration between adjacent layers. As the prepreg tape was 0.35 mm thick and 6 revolutions were performed, a maximum composite thickness of 2.1 mm would be expected in the case of a perfect joining between adjacent layers without interpenetration. However, due to the lack of uniformity of the provided tape, the non perfect circular shape of the first tape revolution, and the tape continuity, an average thickness about 2.6 mm was measured in the case of the composite sample processed at 18 deg/s. In this case, the structure of the superimposed layers was still visible through the thickness of the consolidated sample. By decreasing the winding speed, the layer interpenetration increases and the thickness decreases. The structure of the tape is already not visible at the winding speed of 16.2 deg/s, but the excess of the matrix flow is revealed by the emergence of the glass fibers from the sample rims (Figure 7b). Even if the fiber protrusion does not affect the mechanical performances, that occurrence should be avoided during processing. Probably the best processing strategy could request the modulation of the consolidation force in dependence of the winding speed.

4. CONCLUSIONS

The laser-assisted consolidation of glass reinforced polypropylene tapes by means of a diode laser is a feasible process. A very low laser power is necessary, in combination with a low consolidation force. By changing the winding speed, a different layer interpenetration is generated into the samples with consequences for the thickness and the mechanical properties of the composites. At low winding speeds, even if a good agglomeration is observed, the process is less reliable due to the very low viscosity of the PP matrix. The application of diode lasers to the winding process of thermoplastic matrix composite could to be optimal due to the combination of several advantages: large spot, low specific power, high radiation absorption. As a result, wide areas can be processed with low time and energy consumption and without generating any material degradation. However, many more tests are necessary to reach a process optimum, and the winding system should be improved. In fact, the obtained reliability is not yet acceptable for any industrial application. In the future development, a particular effort will be made to reduce the large thickness variation, and different cylinders with different fiber orientations will be produced, tested and compared with other manufacturing techniques. This way, the proposed technology could be used to produce typical filament wound structures such as pressure vessels or structural tubes.

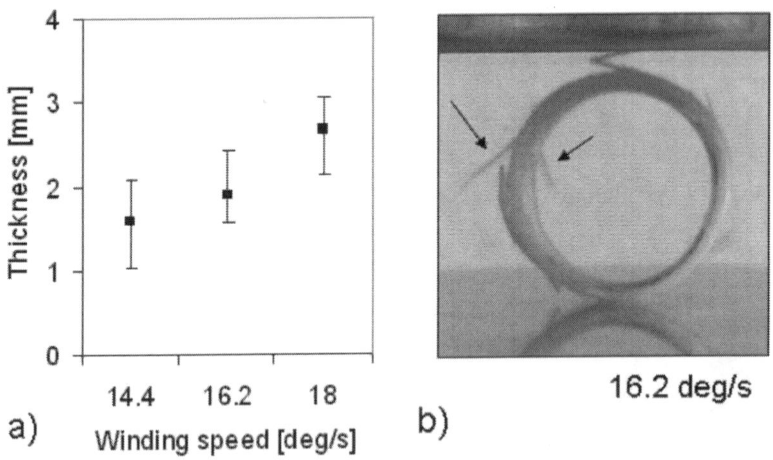

Figure 7. (a) Effect of the winding speed on the composite thickness and (b) appearance of a sample wound at 16.2 deg/s.

ACKNOWLEDGEMENTS

The authors are grateful to Gaetano Spadoni for the support in the experimentation.

REFERENCES AND NOTES

1. Agarwal, V.; Guceri, S.I.; McCullough, R.L.; Schultz, J.M. Thermal characterization of the laser-assisted consolidation process. *J. Thermoplast. Compos. Mater.* **1992**, *5*, 115–135.

2. Rosselli, F.; Santare, M.H.; Güçeri, S.I. Effects of processing on laser assisted thermoplastic tape consolidation.*Composites Part A* **1997**, *28*, 1023–1033.

3. Pistor, C.M.; Yardimci, M.A.; Güçeri, S.I. On-line consolidation of thermoplastic composites using laser scanning.*Composites Part A* **1999**, *30*, 1149–1157.

4. Wang, X.; Lou, D.Y.S. Simple thermal model for online laser curing of thermoset composites in filament winding. *Int. J. Energy Res.* **2003**, *27*, 377–388.

5. Torres, F.G.; Bush, S.F. Sheet extrusion and thermoforming of discrete long glass fibre reinforced polypropylene.*Composites Part A* **2000**, *31*, 1289–1294.

6. Zampaloni, M.A.; Pourboghrat, F.; Yu, W.-R. Stamp thermo-hydroforming: a new method for processing fiber-reinforced thermoplastic composite sheets. *J. Thermoplast. Compos. Mater.* **2004**, *17*, 31–50.

7. Schlottermüller, M.; Lü, H.; Roth, Y.; Himmel, N.; Schledjewski, R.; Mitschang, P. Thermal residual stress simulation in thermoplastic filament winding process. *J. Thermoplast. Compos. Mater.* **2003**, *16*, 497–519.

8. Lü, H.; Schlottermüller, M.; Himmel, N.; Schledjewski, R. Effects of tape tension on residual stress in thermoplastic composite filament winding. *J. Thermoplast. Compos. Mater.* **2005**, *18*, 469–487.

9. Toso, Y.M.P.; Ermanni, P.; Poulikakos, D. Thermal phenomena in fiber-reinforced thermoplastic tape winding process: computational simulations and experimental validations. *J. Compos. Mater.* **2004**, *38*, 107–135.

10. Antunes, P.J.; Dias, G.R.; Nunes, J.P.; van Hattum, F.W.J.; Oliveira, T. Finite element modelling of thermoplastic matrix composite gas cylinders. *J. Thermoplast. Compos. Mater.* **2008**, *21*, 411–441.

11. Squeo, E.A.; Prosperi, C.; Quadrini, F. Forming of long glass fiber reinforced polypropylene. In Proceedings of The 2nd International Conference on Polymers Processing in Engineering PPE 2009, Galati, Romania, 22–23 October 2009.

11

STEADY-STATE BEHAVIOR OF SEMICONDUCTOR LASER DIODES SUBJECT TO ARBITRARY LEVELS OF EXTERNAL OPTICAL FEEDBACK

Qin Zou

Institut Mines-Telecom, Telecom SudParis, Dépatement Electronique et Physique, UMR 5157 CNRS, Evry, France and CEA Saclay Nano-Innov, Gif sur Yvette, France

ABSTRACT

This paper investigates the steady-state behavior of a semiconductor laser subject to arbitrary levels of external optical feedback by means of an iterative travelling-wave (ITW) model. Analytical expressions are developed based on an iterative equation. We show that, as in good agreement with previous work, in the weak-feedback regime of operation except for a phase shift the ITW model will be simplified to the Lang-Kobayashi (LK) model, and that in the case where this phase shift is equal to zero the ITW model is identical to the LK model. The present work is of use in particular for distinguishing the coherence-collapse regime from the strong-feedback regime where low-intensity-noise and narrow-linewidth laser operation would be possible at high feedback levels with re-stabilization of the compound laser system.

KEYWORDS

Semiconductor Lasers; External Optical Feedback; Iterative Traveling-Wave Model; Compound Cavity Modes; Lang-Kobayashi Model; External Cavity Modes; Feedback Lasers

1. INTRODUCTION

Continuous efforts have been made since the last three decades on the dynamics of semiconductor lasers with external optical feedback, because of a large variety of interesting properties they exhibit. A laser with delayed feedback builds an ideal system for analyzing and exploring typical phenomena encountered in a nonlinear time-delayed system, such as bifurcations, thresholds of instability (or stability) and routes to deterministic chaos. On the other hand, external optical feedback can severely affect the spectral behavior of a laser. It can also produce undesirable effects whose effective control is therefore essential for many applications such as optical-fiberbased transmission and sensor systems, since both of them are highly dependent on the spectral quality (temporal coherence and frequency stability) of the used light sources.

In parallel with numerous experimental investigations, theoretical approaches have also been developed aimed at a better understanding of the nonlinear dynamics of a compound laser system (see for example, Tromborg et al. [1], Schunk and Petermann [2], and Binder and Cormack [3]). Most of these approaches have been developed on the basis of the rate equations proposed by Lang and Kobayashi (LK) [4]. For the weak-feedback regime of operation (feedback power ratio less than −30 dB), these approaches have been found to describe adequately various phenomena so far observed experimentally, such as the threshold of coherence collapse (CC), the low-frequency fluctuations (LFFs), and the period-doubling route to chaos [5]. The LK rate equations have also been used to explain the physical mechanisms of the steady-state [6] and transient [7] LFFs, as well as of the chaotic itinerancy for the case of relatively strong feedback [6].

It has been shown that the LK rate equations can be solved analytically by use of asymptotic methods (A detailed description can be found in [8]). In this approach, a laser with weak optical feedback is regarded as a weaklyperturbed nonlinear dynamic system and the threshold of instability corresponds to the first Hopf bifurcation of the LK rate equations. An attempt has been made at interpreting experimental findings of InAs/InP quantum-dash Fabry-Perot lasers by using this approach, such as the onset of CC and the transition from the regime of LFFs to the regime of so-termed fully-developed coherence collapse (FDCC) [9].

Originally, the LK rate equations were proposed to model a single-mode laser with weak feedback and large delays. When the reflectivity of the external reflecting surface is comparable with or greater than the laser facet reflectivity, strong feedback should be taken into account. In this case, the use of the LK model would no longer be justified. Thus, in order to describe the behavior of a feedback laser with arbitrary feedback levels, an iterative traveling-wave (ITW) model was developed [10,11]. By using this model, dynamic and noise properties of a laser subject to strong optical feedback were numerically investigated [12]. The ITW model predicts in particular a significant decrease of the intensity noise in the strongfeedback regime.

More recently, Radziunas et al. used the travelingwave (TW) approach proposed in [13,14] to model a feedback laser, where the system is described by partial differential equations for the electrical fields which counter-propagate along the longitudinal axis of the laser and are coupled through the usual carrier rate equation. A comparison has been made between the LK and TW models, with emphasis on the stability analysis of cavity modes in their continuous-wave states [15].

This paper investigates the steady-state behavior of a feedback laser with arbitrary feedback levels by means of the ITW model. It may be considered as an extension of the works of Langley et al. [12] and Spencer et al. [16]. We provide additional information about the physical insight into a compound laser system and discuss the similarities and the differences between the ITW and LK models. In Section 2, steady-state solutions will be derived for the external cavity modes and compared with previous work. In Sections 3 and 4, a detailed quantitative comparison between the ITW and LK models will be made and the rigorous condition will be given, under which the ITW model will be simplified to the LK model. Finally, Section 5 will summarize our conclusions.

2. ITERATIVE TRAVELING-WAVE MODEL

Consider the configuration of **Figure 1**. A single-longitudinal-mode laser diode is in resonance with an external Fabry-Perot cavity. We assume that r_1, r_2 and r_3 are all real and dispersionless. For this three-mirror system, the dominant resonator is defined by the mirrors with reflection coefficients $r_1$and r_3, and multiple round trips inside the external cavity should be in general taken into account for an arbitrary feedback level.

2.1. Steady-State Solutions

The right-moving electrical field $\tilde{E}(t)$, calculated at steps of the internal round-trip time τ_{in} (in seconds) satisfies the following iterative equation [12,16]

$$\tilde{E}(t+\tau_{in}) = \frac{1}{r_2^2}\exp\left\{\frac{\tau_{in}G_N}{2}(1+j\alpha)\left[N(t)-N_{th}\right]\right\} \times\left[\tilde{E}(t)-\left(1-r_2^2\right)\sum_{k=0}^{M}\left(-r_2r_3\right)^k\times\tilde{E}(t-k\tau)\exp\left(-jk\Delta_0\right)\right] \quad (1)$$

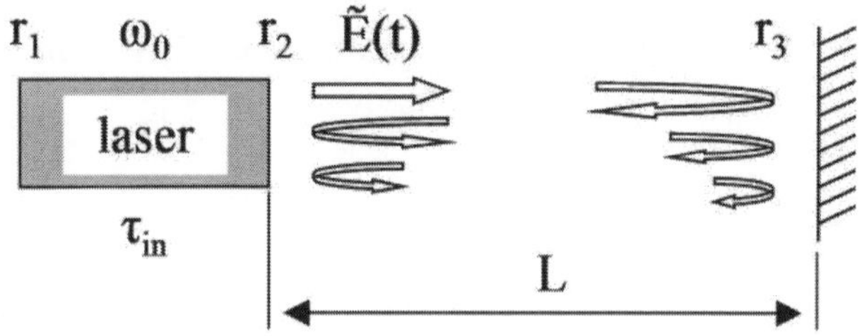

Figure 1. Schematic drawing of a laser diode with external optical feedback. ω_0: emission (angular) frequency of the laser without feedback; τ_{in}: internal round-trip time; r_1: reflection coefficient of the rear facet of the laser; r_2: reflection coefficient of the front facet of the laser; r_3: reflection coefficient of the external mirror; $\tilde{E}(t)$: right-moving electric field passing through the laser front facet; L: length of external cavity assumed empty.

In this equation, G_N (in $s^{-1}\cdot m^3$) is the differential gain; α is the linewidth enhancement factor (LEF); $N(t)$ (in m^{-3}) is the carrier density and N_{th} its threshold; τ (in seconds) is the external round-trip time and Δ_0 (in rad, $\Delta_0 = \omega_0\tau$) is the initial feedback phase associated with the emission frequency of the solitary laser operating near threshold.

By inserting $\tilde{E}(t)=\tilde{E}_0(t)\exp(j\omega t)$ into Equation (1) and considering steady-state solutions, we obtain the following expressions for the excess gain ΔG (in s^{-1}) and the feedback phase Δ (in rad, $\Delta = \omega\tau$ with ω: possible emission frequency) due to the compound structure

$$\Delta G = \frac{-1}{\tau_{in}} \ln \frac{(1-D)^2 + E^2}{r_2^4} \tag{2}$$

and

$$\mathrm{tg}\left(b_2 - \Delta \frac{\tau_{in}}{\tau}\right) + \frac{E}{1-D} = 0. \tag{3}$$

In the above two equations, D (dimensionless), E (dimensionless) and b_2 (in rad) are written respectively as

$$D = \left(1 - r_2^2\right) \sum_{k=0}^{M} \left(-r_2 r_3\right)^k \cos\left(k\Omega\right) \tag{4}$$

$$E = \left(1 - r_2^2\right) \sum_{k=0}^{M} \left(-r_2 r_3\right)^k \sin\left(k\Omega\right) \tag{5}$$

and

$$b_2 = \frac{\tau_{in}\alpha}{2} \Delta G = \frac{-\alpha}{2} \ln \frac{(1-D)^2 + E^2}{r_2^4} \tag{6}$$

where $\Omega = \Delta_0 + \Delta$.

The steady-state behavior of a possible mode (with phase Δ and excess gain ΔG) produced by a feedback laser under multiple-reflection configuration is described by Equations (2)-(6). We will show in Section 3 that in the case of low feedback levels $(r_3 \ll 1)$, except for a phase shift, these equations will reduce to the well-employed forms obtained from the LK rate equations. In the following, a cavity mode referred to the LK model will be called an external cavity mode (ECM) and if this mode is referred to the ITW model, it will be called a compound cavity mode (CCM), as suggested in [15].

2.2. Initial Feedback Phase

Let us first examine the initial feedback phase Δ_0. With a given system, the phase (or normalized emission frequency) Δ of a possible mode, being in dependence on Δ_0, is determined through the so-called phase equation. Under the consideration of low feedback levels, the phase equation has a simpler form and two particular situations have been introduced and widely studied, where Δ is predefined and Δ_0 is then determined from Δ. The first situation corresponds to the "maximum gain mode" defined by the condition $\Delta = 0(\mathrm{mod}\, 2\pi)$. This gives rise to $\Delta_0 = \gamma\tau\alpha(\mathrm{mod}\, 2\pi)$, where γ (in s^{-1}) is the feedback rate defined as usual by [1,2]

$$\gamma = \frac{\left(1 - r_2^2\right) r_3}{r_2 \tau_{in}}. \tag{7}$$

In the second situation, the initial frequency remains unchanged $(\Lambda = \Lambda_0)$ and the related mode is called the "minimum linewidth mode". We have thus $\Delta_0 = -\mathrm{tg}^{-1}(\alpha)$.

We introduce here another category of modes, where Δ_0 is determined directly from Equation (3). So by putting $r_3 = 0$ and M = 0 in Equations (4) and (5), we obtain $D = 1 - r_2^2$, E = 0 and, from Equation (6), $b_2 = 0$. Equation (3) becomes then

$$\Delta_0 = \frac{\tau}{\tau_{in}} m\pi \left(m = 0, \pm 1, \pm 2, \cdots\right). \tag{8}$$

It follows that the change of Δ_0 values is parameterized by the ratio τ/τ_{in}. We note that for both the "maximum gain mode" and the "minimum linewidth mode" there exists the solution $\Delta_0 = 0$ (when $\alpha = 0$), and that this corresponds to the zero-order solution (m = 0) of Equation (8). For simplicity and principle demonstration, we will use in the following $\Delta_0 = 0$ as the value of the initial feedback phase. We will show that in the weak-feedback regime $(r_3 \ll 1)$ quantitative agreement between the ITW and LK models can be obtained only with this initial phase value.

2.3. Excess Gain

A possible CCM has its gain written as

$$G = \tau_p^{-1} + \Delta G \tag{9}$$

where τ_p (in seconds) is the photon lifetime and ΔG is the excess gain due to feedback whose expression is given by Equation (2). A contour plot of the evolution of ΔG as functions of Δ and $r_3$is shown in **Figure 2**. As can be seen from this figure, ΔG peaks at the critical point $r_3 = r_2$, which corresponds to a symmetrical (FabryPerot) external cavity.

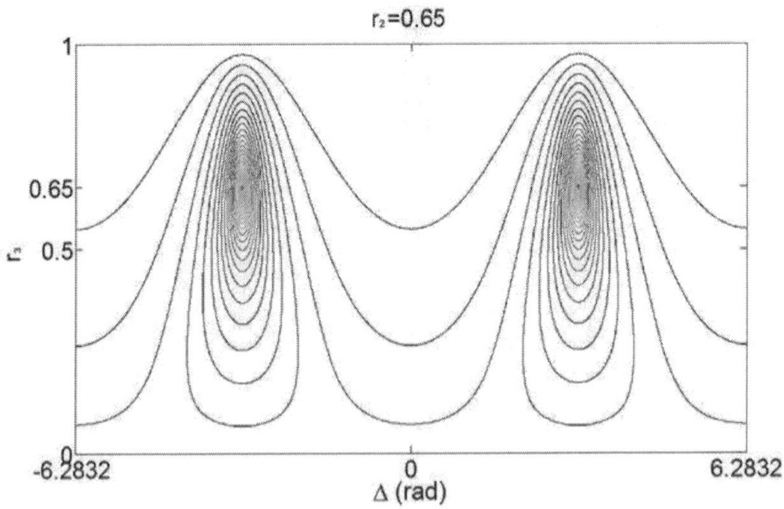

Figure 2. Contour plot of the excess gain ΔG in the vicinity of the initial feedback phase Δ_0 ($\Delta_0 = 0$), as functions of the CCM feedback phase Δ and the reflection coefficient r_3, for $r_2 = 0.65$, $\alpha = 5$, $\tau_{in} = 9$ ps, $\tau = 0.9$ ns, and M = 100.

2.4. Modal Density of Photons

From the standard rate equation for the carrier density N [1,4] and by assuming a linear relation between the gain G and N, we can derive the expression for the density of photons (or photon number) I (in m^{-3}) of a CCM. Here I is calculated directly from $\tilde{E}(t)$, we have thus in steady state

$$I = \left|\tilde{E}_0\right|^2 = \frac{I_s - \frac{\tau_p}{\tau_s}\frac{\Delta G}{G_N}}{1+\tau_p \Delta G} \tag{10}$$

where I_s is the density of photons for the solitary laser. It is written as

$$I_s = \tau_p\left(J - \frac{N_{th}}{\tau_s}\right) \tag{11}$$

where J (in $s^{-1}\cdot m^{-3}$) is the pumping current and τ_s (in seconds) is the carrier life time. An example of the typical evolution of I as functions of Δ and r_3 is shown in **Figure 3**.

A typical phenomenon can be observed from this figure: high values of the reflection coefficient $r_3$do have an effect on enhancement of the density of photons. In this example, I is much higher in the strong-feedback regime than in the moderate-feedback regime (~5 × 10^{20} m^{-3} against ~3 × 10^{20} m^{-3}, which is the threshold value I_s for the solitary laser). As a consequence, the average intensity noise would be decreased in the strong-feedback regime because of higher I values. This result agrees quite well with the work of Langley et al. [12]. They used the ITW model to characterize the transition from the CC regime to the strong-feedback regime for a given r_2 value through the relative intensity noise (RIN) (Figure 2(a), in [12]). We see from this figure that with increasing of r_3, there are three well distinguished regimes for the RIN: the weak-feedback regime (r_3 < 0.005), the noisy CC regime, and the strong-feedback regime (r_3 >

0.1). The RIN increases in the CC regime as expected but decreases significantly (more than 10 dB/Hz) in the strong-feedback regime compared to its values in the weak-feedback regime. This result implies that a stable laser operation with low intensity-noise levels would be possible under the condition of strong feedback. In fact, stable and narrow-linewidth operation has been already observed with systems in configuration of strong optical feedback [17].

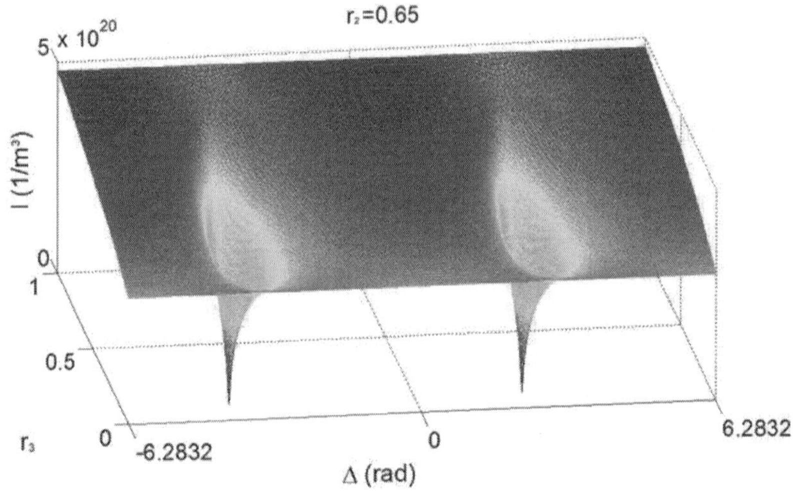

Figure 3. Plot of the density of photons I in the vicinity of the initial feedback phase Δ_0 ($\Delta_0 = 0$) as functions of the CCM feedback phase Δ ranging from -2π to 2π and the reflection coefficient r_3, for $r_2 = 0.65$. The values of the other parameters are $\alpha = 5$, $\tau_{in} = 9$ ps, $\tau = 0.9$ ns, $\tau_s = 2$ ns, $\tau_p = 2$ ps, $G_N = 1.1 \times 10^{-12}$ $s^{-1}\cdot m^3$, $I_s = 3 \times 10^{20}$ m^{-3}, and M = 100.

Another phenomenon is related to the FDCC regime [7]. This regime has been identified for a large pumping current, corresponding therefore to higher output power. In a recent investigation, InAs/InP quantum-dash FabryPerot lasers emitting at 1.57 μm were assessed for their tolerance to external optical feedback by using a freespace setup with a "short" (L = 0.5 m) external cavity [18]. In these experiments, the regime of FDCC was attained for a pumping current of about 30 mA at a rather high feedback level (−1.2 dB in terms of power ratio) with a 600-μm-long laser. As can be seen from the measured RF (radio frequency) spectra reported in this reference ([18], **Figure 5**(b)), the FDCC regime is characterized by a significant increase of the RF peak power around the relaxation oscillation frequency of the solitary laser, and hence by a high output power in steady state at high levels of feedback. We note that such a behavior manifested by a coherence-collapsed laser can be quite well understood using the formalism developed from the ITW model as can be seen from the RIN spectra simulations ([12], **Figure 2**(b)), and that the FDCC regime cor responds roughly to the beginning of the strong-feedback regime, as can be seen from the plot of photon density, i.e. **Figure 3**, in the present text.

2.5. Phase Condition

For a given compound-cavity structure, the phase condition determines the emission frequency of a possible CCM. The phase Δ associated with this mode should satisfy the transcendental Equation (3). In general only numerical solutions are possible.

An example of bifurcation diagrams of the CCMs is shown in **Figure 5**. The value of each point was obtained by a numerical solution of Equation (3). For greater $|\Delta|$ values, it will suffice to repeat the "pattern". We present in **Figure 4** two particular regions in **Figure 5**, where is found in (a) the shape predicted by the LK model as expected. We see from **Figure 5** that, when $r_3 < r_2$, same as a classical bifurcation pattern, all the modes (except the first mode Δ_0) emerge by pairs and their number progressively increases with r_3. The maximum number of modes is attained at the point $r_3 = r_2$, corresponding to a symmetrical external cavity. For $r_3 > r_2$, the modes will disappear also by pairs. Finally, the number of modes will become minimal in the strong-feedback regime. We think that **Figure 5** is equivalent to the bifurcation diagram for the normalized carrier density illustrated in [12] (**Figure 4**), showing clearly that high feedback levels can prevent a feedback laser from noisy output.

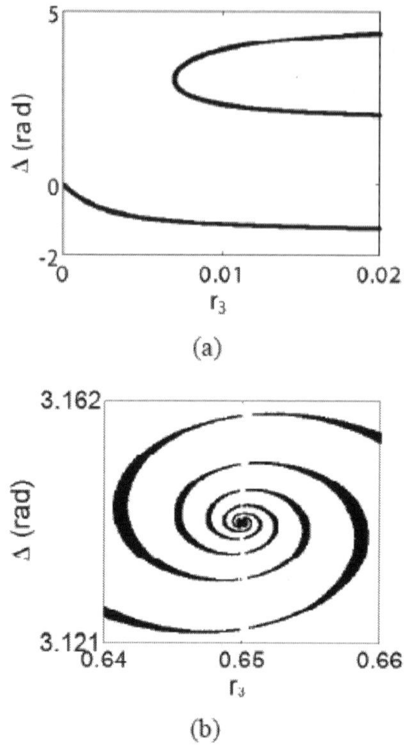

Figure 4. Zoom on two particular regions in **Figure 5**. (a) Region corresponding to small values of r_3 (moderatefeedback regime, where the use of the LK model may still be justified); (b) Whirl-shape region near the critical point $r_3 = r_2$, at which the maximum number of CCMs is found.

3. CONVERGENCE TO THE MODEL OF LANG AND KOBAYASHI

In this section, we will show that in the weak-feedback regime $(r_3 \ll 1)$, except for a phase term, the ITW model will reduce to the LK model.

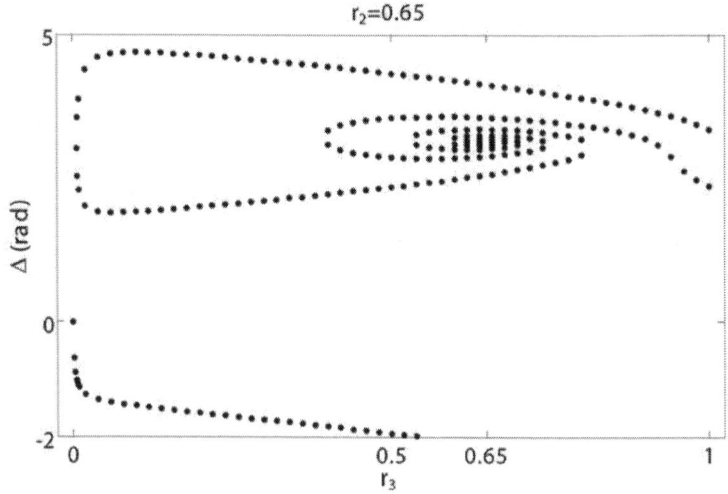

Figure 5. Bifurcation diagram of the CCMs. The values of the parameters are $\alpha = 5$, $\tau_{in} = 9$ ps, $\tau = 0.9$ ns, $\Delta_0 = 0$, $r_2 = 0.65$, and M = 100.

Thus, in the case of low feedback levels, there will be only one reflection (M = 1). Equations (4) and (5) are simplified as

$$D = \left(1 - r_2^2\right)\left[1 - r_2 r_3 \cos(\Omega)\right] \tag{12}$$

$$E = -\left(1 - r_2^2\right) r_2 r_3 \sin(\Omega). \tag{13}$$

Inserting the above two equations into Equation (2) and using the approximations $\left(1 - r_2^2\right) r_3 / r_2 \ll 1$ and $\ln(1 + x) \approx x$ for small x values, we obtain the expression for the steady-state excess gain

$$\Delta G = -2\gamma \cos(\Omega). \tag{14}$$

In the same way, for the phase Equation (3), we have

$$\mathrm{tg}\left(b_2 - \Delta\frac{\tau_{in}}{\tau}\right) = \frac{\left(1-r_2^2\right)r_2 r_3 \sin(\Omega)}{r_2^2 + \left(1-r_2^2\right)r_2 r_3 \cos(\Omega)} \quad (15)$$

where

$$b_2 = -\gamma\tau_{in}\alpha\cos(\Omega). \quad (16)$$

By making some approximations and using $\mathrm{tg}(x) \approx x$, we obtain the phase condition

$$\Delta = -\gamma\tau\left[\alpha\cos(\Omega) + \sin(\Omega)\right] \quad (17)$$

It follows that except for a phase shift Δ_0, Equations (14) and (17) have the same forms as those derived from the LK rate equations. This result confirms the work of Radziunas et al. [15]. They showed a good qualitative agreement between the ECM and CCM solutions at moderate feedback levels. They also found a quantitative agreement for low feedback levels.

4. DIAGRAM OF THE PHOTON DENSITY VERSUS THE MODE PHASE

It is known that for a laser operating in the weak-even moderate-feedback regime a common way to represent the possible steady states of the ECMs at a fixed feedback level is through an ellipse showing the density of photons I versus the feedback phase Δ, and that only a finite number of ECM points are possible which are all located on the ellipse [8,19].

For the case of an arbitrary feedback level, the $I-\Delta$ diagram with a given pumping current J can be established first by expressing Δ as a function of ΔG and then by combining the result with Equation (10). We have respectively from Equations (2) and (3)

$$\Delta = \frac{\tau\alpha\Delta G}{2} - \frac{\tau}{\tau_{in}}\mathrm{tg}^{-1}\left(\frac{E}{D-1}\right) \tag{18}$$

and from Equations (14) and (17) for the case of $r_3 \ll 1$

$$\Delta = \frac{\tau\alpha\Delta G}{2} \pm \frac{\tau}{2}\sqrt{4\gamma^2 - \Delta G^2}. \tag{19}$$

Figure 6 shows the plots of the $I-\Delta$ diagram for various values of r_3, using respectively Equations (18) and (19). Two typical phenomena inside the CC regime are clearly shown in this figure: a "banana"-like shape and a shift, to positive phase values, of the CCM fixed points due to the inclusion of multiple reflections in the ITW model, as also observed by Spencer et al. when they used the ITW model to establish the possible steady states in the I versus $\Delta/(2\pi\tau)$ plane ([16],**Figure 1**). We also observe, in **Figure 6**, a perfect overlap of the two ellipses for r_3 = 0.003. This is because we have taken (for simplicity and principle demonstration) $\Delta_0 = 0$ as the initial phase value.

Finally, let us make a direct comparison between the iteration equation for the ITW model and the rate equation for the LK model. It can easily be shown that at low feedback levels, any stationary solution to Equation (1) will lead to the following equation

$$\frac{G_N}{2}(1+j\alpha)(N_s - N_{th}) + \gamma\exp(-j\Omega) - j\omega = 0 \tag{20}$$

where N_s is the steady-state carrier density. Under the same condition, the LK rate equation will reduce to

$$\frac{G_N}{2}(1+j\alpha)(N_s - N_{th})$$
$$+\gamma\exp(-j\Delta) - j\omega + j\Delta_0/\tau = 0. \tag{21}$$

These two equations show, together with Equations (14) and (17), that the two models are strictly identical if and only if $\Delta_0 = 0$ [see also **Figure 6**(b) for $r_3 = 0.003$].

5. CONCLUSIONS

This paper provides additional information about the physical insight into a compound laser system with arbitrary feedback levels. Analytical expressions have been developed based on an iterative travelling-wave model, which enable a characterization in a rigorous way of a cavity mode in its steady state. We show that with decreasing of the reflection coefficient of the external mirror three regimes emerge successively which can clearly be distinguished from bifurcation diagram and gain plot: they are strong-feedback, coherence-collapse, and moderate-feedback regimes. This latter covers the weakfeedback regime where the use of the model of Lang and Kobayashi is entirely justified.

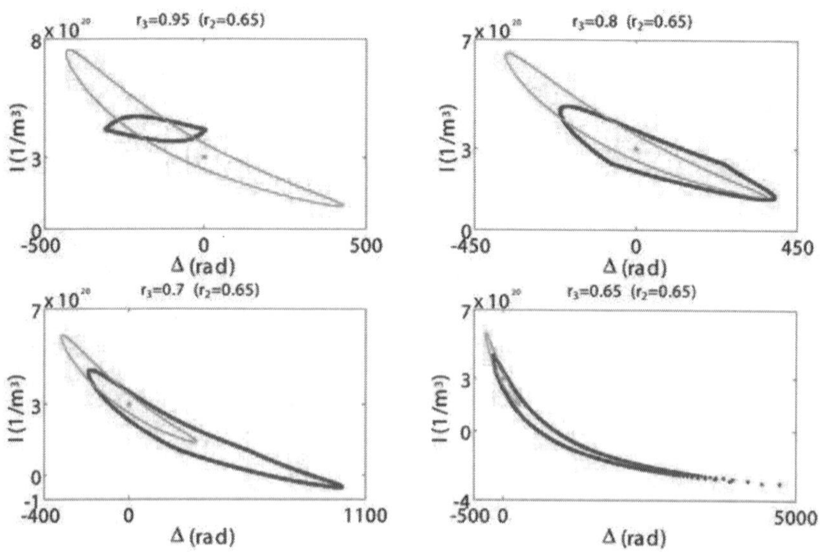

(a)

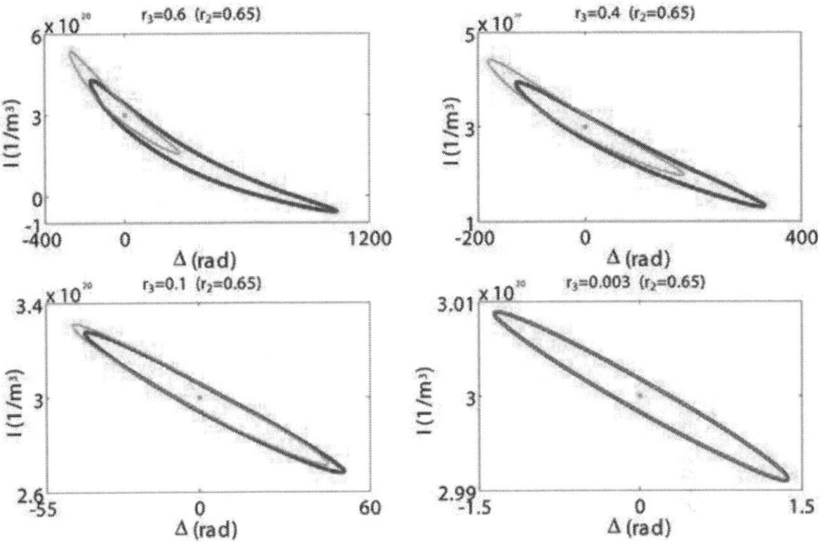

(b)

Figure 6. Evolution of the I – Δ diagram as a function of r_3. The diagrams are determined respectively from Equation (18) (red curves, M = 100) and Equation (19) (green curves). The green dots refer to the solitary-laser solution. (a) $r_3 = 0.95$, 0.8, 0.7 and 0.65; (b) $r_3 = 0.6$, 0.4, 0.1 and 0.003. The values of the other parameters are $r_2 = 0.65$, $\Delta_0 = 0$, $\alpha = 5$, $\tau_{in} = 9$ ps, $\tau = 0.9$ ns, $\tau_s = 2$ ns, $\tau_p = 2$ ps, $G_N = 1.1 \times 10^{-12}\ s^{-1} \cdot m^3$, and $I_s = 3 \times 10^{20}\ m^{-3}$.

We find that maximum number of modes is obtained when the external cavity becomes symmetrical. This state may cause a noisiest laser output. In the strong-feedback regime, a feedback laser is characterized by a minimum mode number and a high density of photons. This behavior confirms previous experimental observations, indicating that beyond the coherence-collapse regime, the system could be re-stabilized and that as a result stable laser operation with low intensity-noise level could be expected with external-mirror reflectivity close to 1. A novel class of modes has been proposed, which is parameterized by the ratio between the external and internal round-trip times. We have also examined the similarities and the differences between the iterative travelling-wave and Lang-Kobayashi models. We find that in the weakfeedback regime these two models are identical only if the initial feedback phase is equal to zero.

The present work is above all useful for determination of the external feedback levels required for stable and noiseless laser output, especially those corresponding to the entrance and the exit of the coherence-collapse regime.

Future investigations will include a detailed analysis of the coherence-collapse regime by means of these two models.

REFERENCES

1. B. Tromborg, J. H. Osmundsen and H. Olesen, "Stability Analysis for a Semiconductor Laser in an External Cavity," IEEE Journal of Quantum Electronics, Vol. 20, No. 9, 1984, pp. 1023-1032.

2. N. Schunk and K. Petermann, "Numerical Analysis of the Feedback Regimes for a Single-Mode Semiconductor Laser with External Feedback," IEEE Journal of Quantum Electronics, Vol. 24, No. 7, 1988, pp. 1242-1247.

3. J. O. Binder and G. D. Cormack, "Mode Selection and Stability of a Semiconductor Laser with Weak Optical Feedback," IEEE Journal of Quantum Electronics, Vol. 25, No. 11, 1989, pp. 2255-2259.

4. R. Lang and K. Kobayashi, "External Optical Feedback Effects on Semiconductor Injection Laser Properties," IEEE Journal of Quantum Electronics, Vol. 16, No. 3, 1980, pp. 347-355.

5. J. Ye, H. Li, and J. G. McInerney, "Period-Doubling Route to Chaos in a Semiconductor Laser with Weak Optical Feedback," Physical Review A, Vol. 47, No. 3, 1993, pp. 2249-2252.

6. T. Sano, "Antimode Dynamics and Chaotic Itinerancy in the Coherence Collapse of Semiconductor Lasers with Optical Feedback," Physical Review A, Vol. 50, No. 3, 1994, pp. 2719-2726.

7. J. Zamora-Munt, C. Masoller and J. García-Ojalvo, "Transient Low-Frequency Fluctuations in Semiconductor Lasers with Optical Feedback," Physical Review A, Vol. 81, No. 3, 2010, Article ID: 033820.

8. T. Erneux and P. Glorieux, "Laser Dynamics," Cambridge University Press, New York, 2010.

9. Q. Zou and S. Azouigui, "Analysis of Coherence-Collapse Regime of Semiconductor Lasers under External Optical Feedback by Perturbation Method," Chapter 5, Semiconductor Laser Diode Technology and Applications, Edition InTech, 2012, pp. 71-86.

10. J. Mørk, "Rep. S48," Danish Center for Applied Mathematics and Mechanics, 1989.

11. F. Sporleder, "Travelling Wave Line Model for Laser Diodes with External Optical Feedback," Proceedings of the URSI International Symposium on Electromagnetic Theory, International Union of Radio Science, Brussels, 1983, pp. 585-588.

12. L. N. Langley, K. A. Shore and J. Mørk, "Dynamical and Noise Properties of Laser Diodes Subject to Strong Optical Feedback," Optics Letters, Vol. 19, No. 24, 1994, pp. 2137-2139.

13. J. E. Carroll, J. Whiteaway and R. Plumb, "Distributed Feedback Semiconductor Lasers," Institution of Electrical Engineers, London and SPIE Optical Engineering Press, 1998.

14. U. Bandelow, M. Radziunas, J. Sieber and M. Wolfrum, "Impact of Gain Dispersion on the Spatio-Temporal Dynamics of

Multisection Lasers," IEEE Journal of Quantum Electronics, Vol. 37, No. 2, 2001, pp. 183-188.

15. M. Radziunas, H. J. Wünsche, B. Krauskopf and M. Wolfrum, "External Cavity Modes in Lang-Kobayashi and Traveling Wave Models," Proceedings of SPIE, Vol. 6184, 2006.

16. P. S. Spencer, C. R. Mirasso and K. A. Shore, "Effect of Strong Optical Feedback on Vertical-Cavity SurfaceEmitting Lasers," IEEE Photonics Technology Letters, Vol. 10, No. 2, 1998, pp. 191-193.

17. C. E. Weiman and L. Holberg, "Using Diode Lasers for Atomic Physics," Review of Scientific Instruments, Vol. 62, No. 1, 1991.

18. S. Azouigui, B. Kelleher, S. P. Hegarty, G. Huyet, B. Dagens, F. Lelarge, A. Accard, D. Make, O. Le Gouezigou, K. Merghem, A. Martinez, Q. Zou and A. Ramdane, "Coherence Collapse and Low-Frequency Fluctuations in Quantum-Dash Based Lasers Emitting at 1.57 μm," Optics Express, Vol. 15, No. 21, 2007, pp. 14155-14162.

19. C. H. Henry and R. F. Kazarinov, "Instability of Semiconductor Lasers Due to Optical Feedback from Distant Reflectors," IEEE Journal of Quantum Electronics, Vol. 22, No. 2, 1986, pp. 294-301.

12

DIODE LASER DETECTION OF GREENHOUSE GASES IN THE NEAR-INFRARED REGION BY WAVELENGTH MODULATION SPECTROSCOPY: PRESSURE DEPENDENCE OF THE DETECTION SENSITIVITY

Takashi Asakawa [1], Nozomu Kanno [2] and Kenichi Tonokura [3,*]

[1] Department of Chemical System Engineering, The University of Tokyo, 7-3-1 Hongo, Bunkyo-ku, Tokyo 113-8656, Japan

[2] Department of Micro-Nano Systems Engineering, Nagoya University, Furo-cho, Chikusa-ku, Nagoya 464-8603, Japan

[3] Environmental Science Center, The University of Tokyo, 7-3-1 Hongo, Bunkyo-ku, Tokyo 113-0033, Japan

ABSTRACT

We have investigated the pressure dependence of the detection sensitivity of CO_2, N_2O and CH_4 using wavelength modulation spectroscopy (WMS) with distributed feed-back diode lasers in the near infrared region. The spectral line shapes and the background noise of the second harmonics (2*f*) detection of the WMS were analyzed theoretically. We determined the optimum pressure conditions in the detection of CO_2, N_2O and CH_4, by taking into consideration the background noise in the WMS. At the optimum total pressure for the detection of CO_2, N_2O and CH_4, the limits of detection in the present system were determined.

KEYWORDS

gas sensing; wavelength modulation spectroscopy; near-infrared absorption spectroscopy

1. INTRODUCTION

Atmospheric pollution is a problem as global warming is caused by the man-made addition of extra amounts of greenhouse gases such as CO_2, N_2O and CH_4. Current average mixing ratios of the trace components CO_2, N_2O and CH_4 in the atmosphere are ~380, ~1.8 and ~0.32 parts per million by volume (ppmv), respectively. CO_2 is released into the atmosphere when fossil fuels, coal, solid waste and wood products are burned, during cement production, and also when land surface cover is changed by humans. Increased CH_4 levels come from fossil fuels, rice cultivation, animal husbandry, biomass burning and landfills. The main anthropogenic sources of N_2O are agriculture, and industrial sources including adipic and nitric acid production. Combustion of solid waste and fossil fuels also contributes to atmospheric N_2O. The need to detect these trace gases has become increasingly important in recent years, both for controlling industrial processes and for monitoring air quality. Since many trace gases have a significant impact on the environment, the development of techniques for its fast, accurate and sensitive detection is required [1–5].

Laser based spectroscopy provides interesting advantages related to its high selectivity and sensitivity in the detection of trace gases. Semiconductor diode lasers emitted the near infrared (NIR) light have played a central role, being tunable spectroscopic light sources which exhibit a relatively low amplitude noise [6]. In addition, their small size and the relatively low cost have made them particularly suitable for the realization of transportable spectrometers employed for *in-situ*measurements [7,8]. In the NIR region, relatively weak overtone or combination vibrational transitions occur for trace gases relevant to the atmospheric environment. Although line strengths of these transitions are a few orders of magnitude lower than transitions belonging to fundamental vibrational bands, cavity ring down spectroscopy (CRDS) [9] and frequency modulation spectroscopy (FMS) with a single-mode diode laser in the near infrared region have allowed to achieve the detection of low concentrations (< sub ppmv) of trace gases [10]. The use of multi-pass absorption cells [11] further improves the sensitivity, since the effective optical path is increased up to several tens or hundreds of meters. Phase-sensitive techniques, which are the core of the FM techniques, significantly reduces the $1/f$ electronic noise, achieving high detection sensitivity. In most cases the detection sensitivity is only limited by noise introduced by undesired optical fringes and drifting of laser power and detector sensitivity. Wavelength modulation spectroscopy (WMS) [12–16] is a kind of FMS with a modulation frequency lower than the spectral line width of interest. In the case of WMS, the lower modulation frequency allows one to use low-frequency circuits and photo detectors, reducing the complexity and the cost of measurement system. Recently, a high sensitive open path CH_4 analyzer based on WMS detection and Herriott cell design with a 30 m total optical path length has been developed which is commercially available [17].

In the present study, we demonstrate the detection of greenhouse gases such as CO_2, CH_4 and N_2O using WMS. The aim of this work is characterized the optimum condition of the detection of greenhouse gases by NIR-WMS. The optimum pressure for the detection of CO_2, N_2O and CH_4 by WMS is investigated both experimentally and theoretically by taking into consideration background noise in the WMS.

2. EXPERIMENTAL

We used WMS in the near-infrared region for trace gas detection. Figure 1 shows a scheme of our experimental apparatus. All optical elements were properly designed to reduce their dimensions and mechanical instability as much as possible. Three distributed feed-back (DFB) diode lasers (NTT Electronics) with a tuning range of ± 1 nm were used as light sources. The first DFB laser with a center wavelength of 1,572 nm was used to detect CO_2 at the $3\nu_1 + \nu_3$ combination band. The second DFB laser with a center wavelength of 1,515 nm was used to detect the N_2O line at the $3\nu_3$ overtone band. The last DFB laser with a center wavelength of 1,651 nm was used to detect CH_4 at the $2\nu_3$ overtone band. The laser power varied as a function of injection current with a maximum of 20 mW. The wavelengths of these DFB lasers were chosen for strong absorbing transitions and no interference of water.

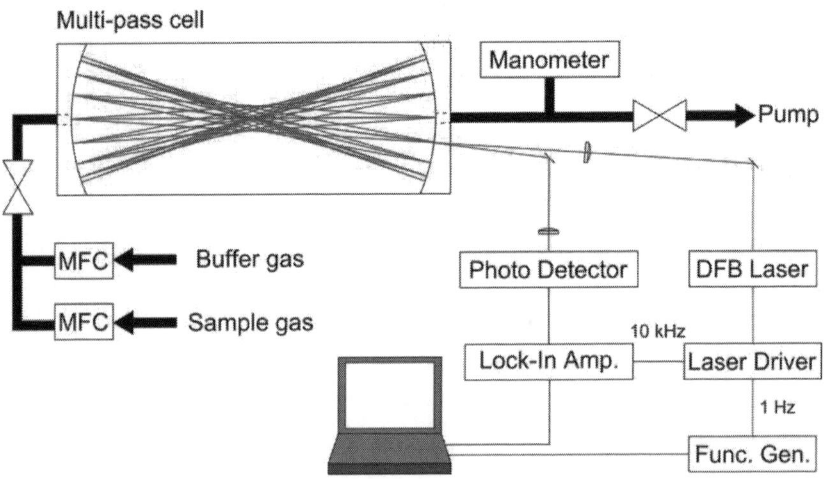

Figure 1. Schematic diagram of the experimental setup (MFC: mass flow controller. Func. Gen.: function generator).

The laser wavelengths were tuned by varying the laser temperatures, while fine tuning was accomplished by changing the laser diode injection currents. To minimize fringe noise from optics, anti-reflection (AR) coated lenses and AR coated and wedged windows were used. The output beam was focused by an AR coated lens (f = 50 cm) to the centre of a Herriott-type multi-pass cell [11] with a distance of 40.4 cm between two mirrors. In the multi-pass cell, the laser beam undergoes multiple reflections between two mirrors; the total number of 74 passes corresponds to the optical path length of 29.91 m. The inner volume of the cell is 900 cm^3. The beam passed through the multi-pass cell was finally focused by a short focal AR coated lens (f = 5 cm) onto an InGaAs photodiode detector (Hamamatsu G5852-11).

The laser wavelength was sinusoidally modulated at 10 kHz, and scanned at 1 Hz around the absorption line. The laser wavelength scan and modulation were performed through a custom-made laser driver. The background bias signal of the laser driver used in this study was same level as that of a commercial driver (e.g., ILX LDC3724C). Under typical conditions, the modulated absorption signal was phase-sensitive detection at twice the modulation frequency (2*f*) using a lock-in amplifier (Stanford Research System SR810 DSP) with the time constant set to 3 ms. The data was acquired via a 16 bit AD PCMCIA card (CONTEC ADA16-8/2(CB)L) to a laptop computer and analyzed using software written in LabVIEW.

N_2 (99.99%) and premixed gases of synthesized air (79% N_2, 21% O_2), 1% CO_2 in N_2, 1% N_2O in N_2, and 0.1% CH_4 in air were purchased from Taiyo Nippon Sanso Co. The flow rate of the feed was controlled with mass flow controllers (Kofloc MODEL3660). The total pressure was monitored with a capacitance manometer (Setra Systems MODEL720). All measurements were performed at room temperature (293 ± 3 K).

3. WMS THEORY

The general theory of the FMS and WMS has been well summarized in the works of previous researchers [13–15]. In the case of sinusoidal modulation with modulation index β and frequency ω_m, the optical field E with carrier frequency ω_0 is given by:

$$E(t) = E_0 \exp[i(\omega_0 t + \beta \sin \omega_m t)] = E_0 \exp(i\omega_0 t) \sum_{n=-\infty}^{+\infty} J_n(\beta) \exp(in\omega_m t) \quad (1)$$

where the expansion in a series of nth-order Bessel functions J_n characterizes the frequency components of the modulated light spectrum. FMS is classified by its modulation frequency ω_m. WMS uses ω_m that is much less than the spectral line width Γ of the absorption line of interest. Conversely, narrowly-defined FMS uses $\omega_m \gg \Gamma$. In the case of WMS, the probe beam modulation is generally treated as instantaneous frequency change:

$$\omega_i(t) \equiv \frac{d}{dt}(\omega_0 t + \beta \sin \omega_m t) = \omega_0 + \Delta F \cos \omega_m t \quad (2)$$

where the maximum frequency deviation from the carrier frequency ω_0 is $\Delta F = \beta\omega_m$. In this model, dispersion effects cannot be considered, and the light intensity transmitted through the sample is expressed by $I_T = I_0 \exp(-\alpha)$, where α is the absorption coefficient.

On the assumption that $\alpha(\omega) \ll 1$, $I_T(\omega) \approx I_0(\omega)[1 - \alpha(\omega)]$, the $\alpha(\omega)$ can be expanded into the Taylor series around the carrier frequency ω_0:

$$I_{\mathrm{T}}(\omega)=I_0\left[1-a(\omega_0)-\left.\frac{\mathrm{d}\alpha}{\mathrm{d}\omega}\right|_{\omega_0}(\omega-\omega_0)-\frac{1}{2!}\left.\frac{\mathrm{d}^2\alpha}{\mathrm{d}\omega^2}\right|_{\omega_0}(\omega-\omega_0)^2\right.$$
$$\left.-\frac{1}{3!}\left.\frac{\mathrm{d}^3\alpha}{\mathrm{d}\omega^3}\right|_{\omega_0}(\omega-\omega_0)^3-\ldots\right]$$
$$=I_0\left[1-a(\omega_0)-\left.\frac{\mathrm{d}\alpha}{\mathrm{d}\omega}\right|_{\omega_0}\Delta F\cos\omega_{\mathrm{m}}t-\frac{1}{2!}\left.\frac{\mathrm{d}^2\alpha}{\mathrm{d}\omega^2}\right|_{\omega_0}(\Delta F)^2\frac{1}{2}(1+\cos 2\omega_{\mathrm{m}}t)\right.$$
$$\left.-\frac{1}{3!}\left.\frac{\mathrm{d}^3\alpha}{\mathrm{d}\omega^3}\right|_{\omega_0}(\Delta F)^3\frac{1}{4}(3\cos\omega_{\mathrm{m}}t+\cos 3\omega_{\mathrm{m}}t)\ldots\right]. \quad (3)$$

Here $\cos^2 \omega_{\mathrm{m}}t$, $\cos^3 \omega_{\mathrm{m}}t$, *etc.*, have been expanded with trigonometric identities for convenience in picking off the power at any frequency of interest. Because the instantaneous frequency varies sinusoidally, the Taylor series of $I_{\mathrm{T}}(\omega)$ is more or less automatically the Fourier series, too. This series will not converge well for some ΔF [16]. However, if the frequency deviation ΔF is sufficiently small compared with Γ, higher-order terms can be neglected, and the WMS signal that is coherently detected at the frequency $n\omega_{\mathrm{m}}$ is proportional to the n-th derivative of the sample absorption.

In the present study, the assumption that $\Delta F \ll \Gamma$ was not valid under some experimental conditions. Thus we used the frequency domain approach to calculate the WMS spectral line shape. In the real system, the DFB laser intensity is not uniform over the FM range, and simultaneous amplitude modulation (AM) of the electric field occurs such that:

$$E(t)=E_0[1+M\sin(\omega_{\mathrm{m}}t+\psi)]\exp[i\omega_0 t+i\beta\sin(\omega_{\mathrm{m}}t)] \quad (4)$$

where M and ψ denote the AM index and the phase difference between AM and FM. By expanding the AM term in term of exponentials, *i.e.*, $M\sin(\omega_{\mathrm{m}}t+\psi) = M/2[\exp\{i(\omega_{\mathrm{m}}t+\psi)\} - \exp\{i(\omega_{\mathrm{m}}t+\psi)\}]$, and replacing the exponential frequency terms as a summation over Bessel functions, one can rewrite Equation (4) as:

$$E(t) = E_0 \exp(i\omega_0 t)\sum_{n=-\infty}^{\infty} r_n \exp(in\omega_m t) \tag{5}$$

where:

$$r_n = \sum_{k=-1}^{1} a_k J_{n-k}(\beta), \qquad a_0 = 1,\ a_{\pm 1} = \frac{\pm M}{2i}\exp(\pm i\psi) \tag{6}$$

Equation (1) is the special case of Equation (5) where $M = 0$.

When the modulated electric field $E(t)$ is passed through a medium exhibiting the field amplitude attenuation $\delta(\omega)$ and the optical phase shift $\varphi(\omega)$, the transmitted optical field is then:

$$E_T(t) = E_0 \exp(i\omega_0 t)\sum_{n=-\infty}^{\infty} \exp(-\delta_n - i\varphi_n) r_n \exp(in\omega_m t) \tag{7}$$

where $\delta_n = \delta(\omega_0 + n\omega_m)$ and $\varphi_n = \varphi(\omega_0 + n\omega_m)$. Because the intensity I equals $c\varepsilon_0 EE^*/2$ in SI units, the absorption coefficient $\alpha$is twice the electric field amplitude attenuation coefficient δ for weak absorption. The transmitted intensity can be expressed as a sum of Fourier components at frequencies corresponding to multiples of ω_m [13]. In the case of $2\omega_m$ detection, the resulting signal is:

$$I = 2I_0[\mathrm{Re}(Z)\cos\theta - \mathrm{Im}(Z)\sin\theta], \quad Z = \sum_{n=-\infty}^{\infty} r_{n+1} r_{n-1}^{*}\left[\exp(-(\delta_{n+1} + \delta_{n-1})) + i(\varphi_{n+1} - \varphi_{n-1})\right] \tag{8}$$

where θ refers to the phase difference between the transmitted intensity and the reference signal in the lock-in amplifier. The in-phase (0 or π) component corresponds to absorption, and the quadrature (±π/2) components correspond to dispersion of the incident light. In the absence of absorption and dispersion, Z has a nonzero value due to residual amplitude modulation (RAM) [14]. In the case of $2\omega_m$ detection, the background bias signal is:

$$I_{\mathrm{RAM}} = \frac{1}{2} M^2 I_0 \cos(\theta + 2\psi + \pi) \tag{9}$$

Since I_{RAM} linearly depends on I_0, low frequency noise in I_0, such as fringe noise from optics and drifting of laser power and detector sensitivity, within the noise equivalent bandwidth of the lock-in amplifier is converted into FM/WM signal background noise.

WM spectra are calculated from Equation (8) assuming that the absorption peak shape is described by the Voigt function. The Voigt function was calculated following the algorithm described by Humlicek [18] and Schreier [19]. The Doppler half width at half maximum (HWHM) was calculated from the temperature and molecular weight. To calculate homogeneous HWHM values, the pressure broadening factor was taken from the HITRAN database [20]. The FM index β was estimated from $\beta = \Delta F/\omega_{\mathrm{m}}$. To calculate the AM index M, the intensity dependence on the laser frequency was measured, separate from the WMS experiments. Table 1 shows the parameters used in the spectral simulation. For the modulation frequency below 750 MHz, the AM-FM phase difference ψ has been found to have a value of ca. $n\pi + \pi/2$ [14]. For the present study, negative dependence of the laser intensity on the laser frequency suggests that $\psi \approx -\pi/2$. In the WMS condition, *i.e.*, $\omega_{\mathrm{m}} \ll \Gamma$, the dispersion term is negligibly smaller than the absorption term, thus the detection phase $\theta \approx 0$ for the maximum signal intensity. In all calculations, the values of ψ and θ were fixed to be $-\pi/2$ and 0, respectively. The signal intensity and noise are defined as peak-to-baseline amplitude and peak-to-peak variation of 2f signal with standard deviation of baseline, as shown in Figure 2.

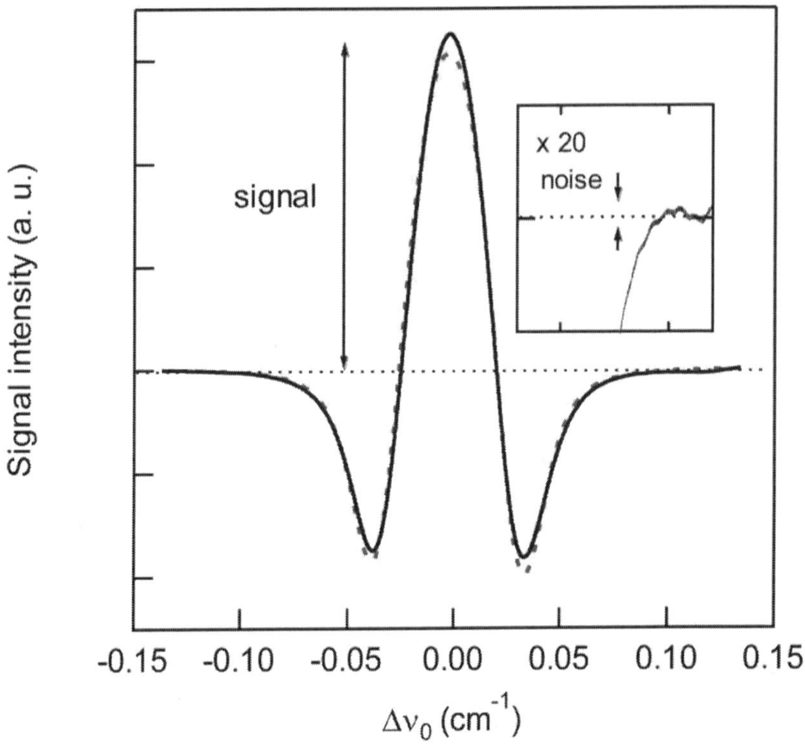

Figure 2. Experimental (solid line) and simulated (dashed line) 2*f* wavelength modulation spectra of 1% CO_2 at around 6,359.97 cm^{-1} and a total pressure of 15 kPa diluted by N_2.

Table 1. Parameters used in the spectral simulation.

Laser	Target	ω_0 [a]	$\Gamma_{Doppler}$ [a]	Γ_{air} [b]	Γ_{self} [b]	$M/\omega_m\beta$ [c]
1	CO_2	6,359.967	0.00593	0.0734	0.1009	0.423
2	N_2O	6,591.437	0.00614	0.0774	0.1009	0.305
3	CH_4	6,046.953	0.00933	0.0627	0.0820	0.288

[a] In units of cm^{-1}. [b] In units of cm^{-1}/atm. [c] In units of 1/cm^{-1}.

4. RESULTS AND DISCUSSION

For the detection of CO_2, we used the $3\nu_1 + \nu_3$ combination band of the R(16) line at the absorption line center of 6,359.97 cm^{-1}, which is free from interference by the other atmospheric species and has a line strength of 1.73×10^{-23} cm^2 $molecule^{-1}cm^{-1}$, according to the HITRAN database [20]. Figure 2 shows experimental and simulated spectra of 1% CO_2 at a total pressure of 15 kPa, diluted by N_2. The experimental spectra were acquired with a lock-in time constant of 3 ms and filter slope of 24 dB/oct, which leads a noise equivalent bandwidth of the low pass filter to be 26 Hz. The simulated spectrum reconstructed the experimental spectrum well. Figure 3 shows the total pressure dependence of signal intensity and background noise. In all experiments, the modulation depth ΔF were adjusted to obtain the maximum signal intensity. The concentration of CO_2 was maintained at 1% in N_2. In the simulation, we assumed that background noise was linearly dependent on M^2, according Equation (9), and the simulated noise intensities were normalized to minimize the residual to experimental values. As can be seen, the simulated results closely agree with the experimental measurements. The signal intensity profile of Figure 3 shows two opposing effects of pressure: decreasing spectral line centre intensities due to pressure broadening and an increasing absorber number density with increasing pressure. At low pressures (Doppler regime), the signal intensity increases linearly with pressure, while at high pressures (Lorentzian regime) the signal intensity become almost constant due to the canceling of two opposing terms with pressure. Figure 4 shows experimental and simulated profiles of the signal to noise ratio (S/N) as a function of total pressure. The optimum total pressure in the detection of CO_2 was found to be around 7 kPa. In the WMS, optimization of the signal intensity occurs at ΔF/HWHM = 2.2 in the Voigt function to calculate the properties of the $2f$ line shape [21].

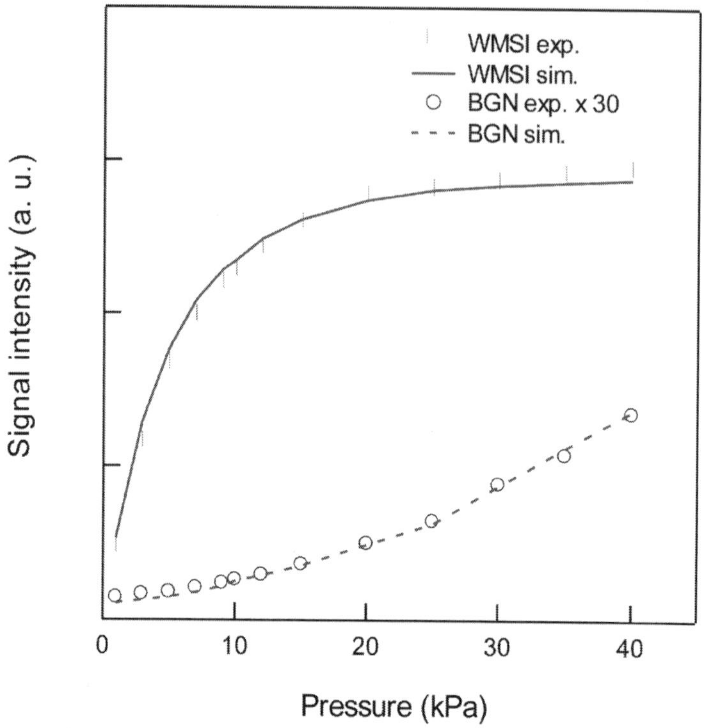

Figure 3. Total pressure dependence of 2*f* wavelength modulation signal intensities (WMSI; •, ———) and background noise (BGN; ○, - - -) for 1% CO_2 at 6,359.97 cm^{-1} diluted by N_2. Circles and lines indicate experimental and simulated results, respectively.

Because the Voigt HWHM is increased by pressure broadening with increasing pressure, the frequency deviation ΔF must be increased in order to maximize the WMS signal with increasing pressure. Since the AM index M depends linearly on ΔF,*i.e.*, $M/\omega_m\beta$ (= $M/\Delta F$) is constant, and the residual amplitude modulation I_{RAM} is linearly dependent on M^2 according toEquation (9), the low frequency noise in I_0 such as the fringe noise from optics and drifting of laser power and detector sensitivity was enlarged by factor of M^2 at higher pressure, as shown in Figure 3.

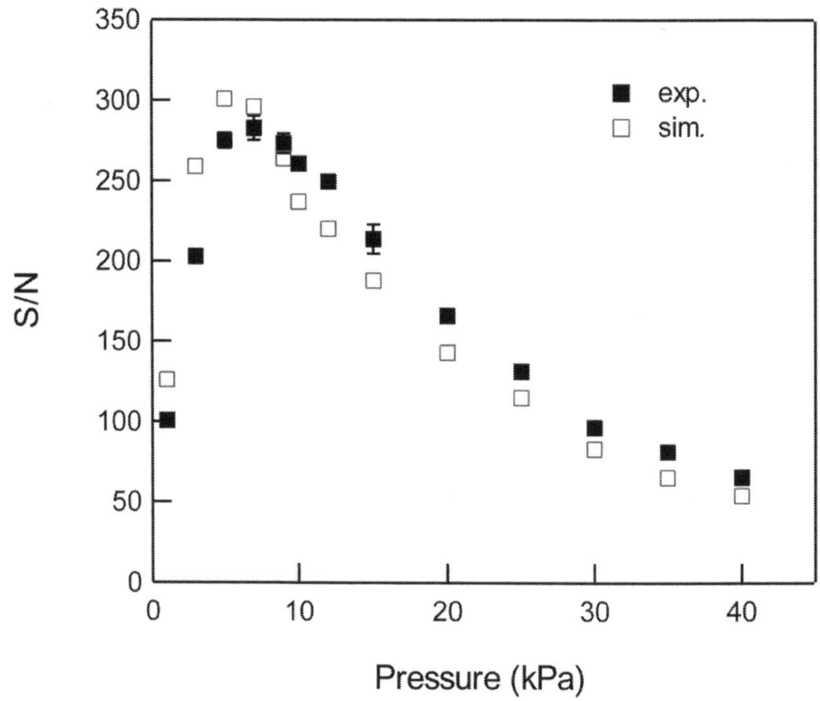

Figure 4. Total pressure dependence of S/N for 1% CO_2 at 6,359.97 cm^{-1} diluted by N_2. Closed and open squares indicate experimental and simulated results, respectively.

In the case of N_2O detection, the $3\nu_3$ overtone band of the R(15) absorption line at 6,591.44 cm^{-1} with a line strength of 2.33×10^{-23} cm^2 $molecule^{-1}$ cm^{-1} was monitored. Figure 5 shows the experimental and simulated spectra of 1% N_2O at a total pressure of 6 kPa. As for the previous experiment of CO_2, the acquired 2*f* WM spectrum of N_2O shown in Figure 5 is in excellent agreement with the simulated spectrum. Figure 6 shows the total pressure dependence of signal intensity and background noise for the 1 % N_2O diluted in N_2. The signal intensity is saturated above 20 kPa. Figure 7 shows experimental and simulated profiles of S/N as a function of total pressure. The simulated results agreed with the experimental results. The optimum total pressure for the detection of N_2O is estimated to be around 6 kPa.

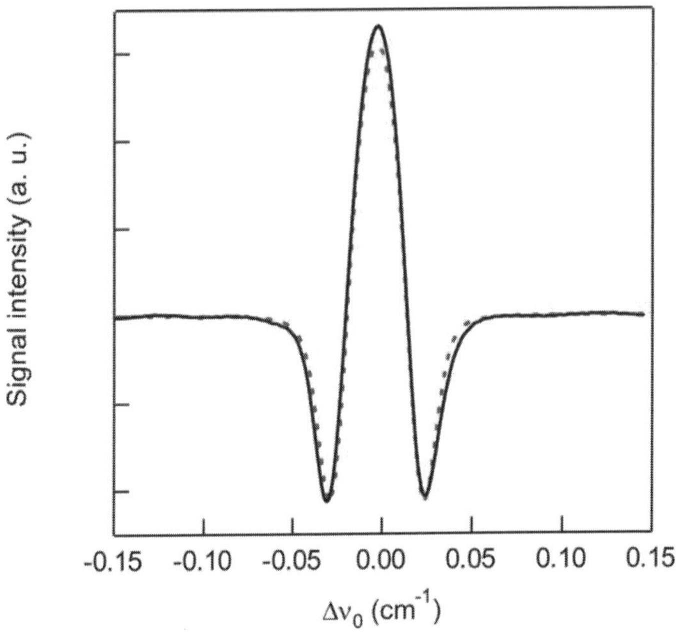

Figure 5. Experimental (solid line) and simulated (dashed line) 2*f* wavelength modulation spectra of 1% N_2O at around 6,591.44 cm^{-1} and a total pressure of 6 kPa diluted by N_2.

For the detection of CH_4, we used the R(3) lines of the $2\nu_3$ overtone band around 6,046.95 cm^{-1} with a line strength of the order 10^{-21} cm^2 $molecule^{-1}$ cm^{-1} [20]. As listed in Table 2, there are three absorption peaks at around 6,046.95 cm^{-1}. Figure 8shows the experimental and simulated spectra of 0.01% CH_4 at around 6,046.95 cm^{-1} and total pressure of 10 kPa. The three absorption lines were considered in the simulation. In the experimental spectra, only one absorption peak was observed at 6,046.95 cm^{-1} due to the overlapping of the three absorptions. Figure 9 shows the total pressure dependence of signal intensity and noise for the 0.01 % CH_4 diluted in N_2. Figure 10 shows the experimental and simulated profiles of S/N as a function of total pressure. The simulated spectrum using the three absorption lines reproduces the experimental results well. The optimum total pressure of CH_4

detection at the 6,046.95 cm^{-1} absorption peak was found to be around 15 kPa. The larger optimum pressure for CH_4 detection compared with those for CO_2 and N_2O would be caused by the broader nature of the absorption peak consisting of three closely separated absorption lines.

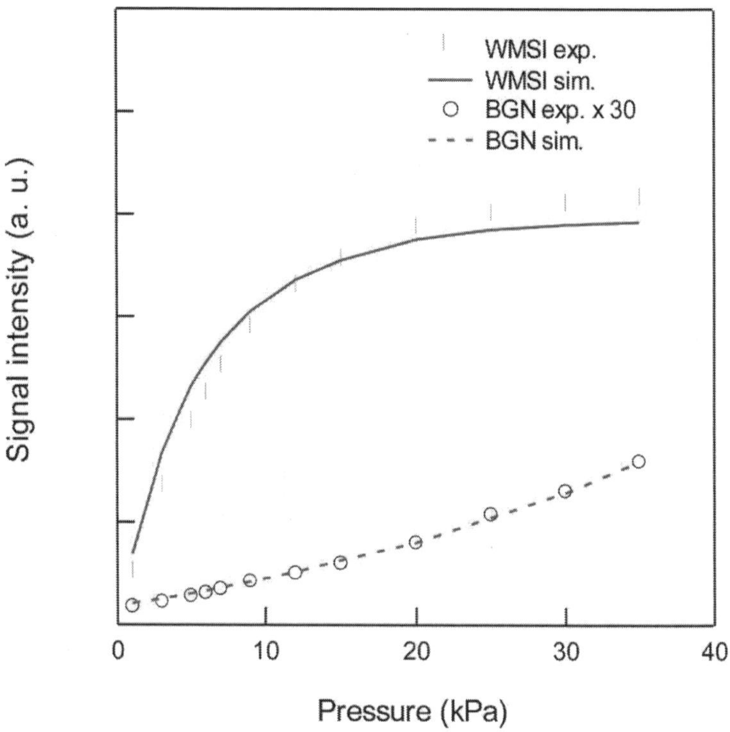

Figure 6. Total pressure dependence of 2*f* wavelength modulation signal intensities (WMSI; •, ———) and back ground noise (BGN; o,— — —) for 1% N_2O at 6,591.44 cm^{-1} diluted by N_2. Circles and lines indicate experimental and simulated results, respectively.

To estimate the limit of detection (LOD) of the present system, we changed the mixing ratio of sample gases, while keeping the optimum value of the total pressure fixed. The measurements were performed with a noise equivalent bandwidth of 26 Hz. From a fit of the signal intensity as a function of CO_2 mixing ratio at 7 kPa, we found, for a signal to noise ratio of 2, the LOD of $[CO_2]^{7kPa}$ = 24 ± 2 ppmv. In the case of N_2O, the LOD at 6 kPa was estimated to be 7.6 ± 0.4 ppmv. For CH_4, we estimated the detection limit and NEAS using the R(3) peaks at 15 kPa to be 60 ± 4 ppbv. The LODs for the three gases had an almost linear dependence on the absorption cross sections. The line strengths of CO_2, N_2O and $CH_4$of the used transitions vary slightly with temperature (0.23, 0.25 and 0.39%, respectively, by temperature change of 1 K around room temperature), however, the LODs are unaffected by temperature change. Parkes *et al.* demonstrated the detection of N_2O by diode laser cavity ring-down spectroscopy using the same $3\nu_3$ overtone band [22]. The limiting sensitivity was 23 ppmv at 1 atm with mirror reflectivities of R = 0.9998, which is comparable with our 2*f* WMS value of 7.6 ppmv. The noise-equivalent absorption sensitivity (NEAS) per scan for a signal to noise ratio of 1 was estimated using following Equation (10):

$$NEAS = \frac{Abs_{min}}{l\sqrt{NEBW/N}} \tag{10}$$

where Abs_{min}, l, $NEBW$ and N denote the absorbance at S/N = 1, the effective optical length of 29.91 m, the noise equivalent bandwidth of 26 Hz and the average number of 20 times, respectively. In the calculation of Abs_{min}, the absorption line shapes were assumed to be described as the Voigt functions. The estimated values of NEAS for CO_2, N_2O and CH_4 were found to be (1.3 ± 0.1) × 10^{-8}, (4.8 ± 0.3) × 10^{-9}, and (5.6 ± 0.4) × 10^{-9} cm^{-1} $Hz^{-1/2}$, respectively. The higher value of NEAS for CO_2 would be caused by the larger value of $M/\omega_m\beta$, as shown in Table 1, which results the larger effect of the residual amplitude modulation.

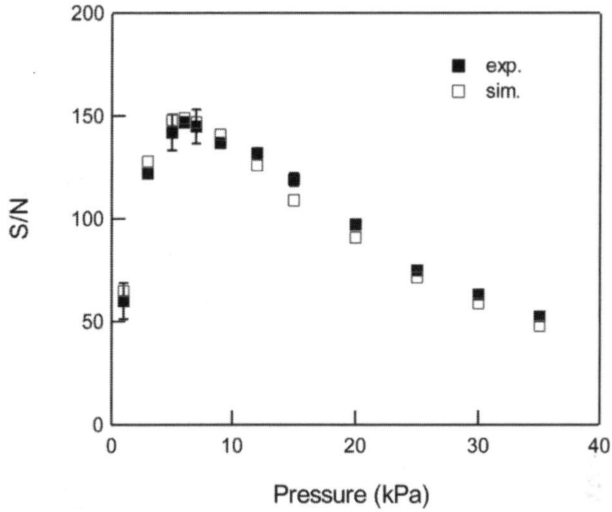

Figure 7. Total pressure dependence of S/N for 1% N_2O at 6,591.44 cm^{-1} diluted by N_2. Closed and open squares indicate experimental and simulated results, respectively.

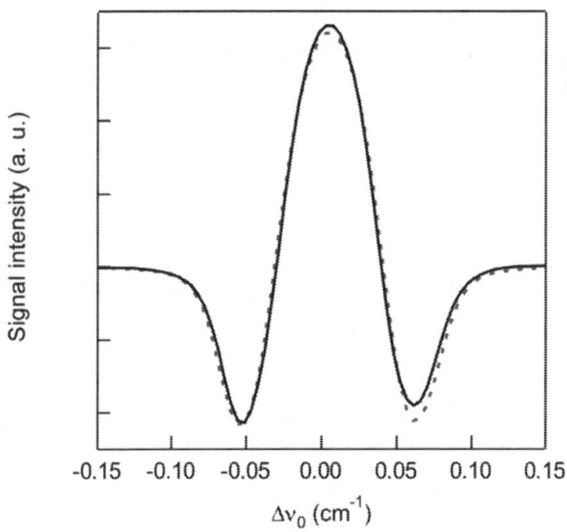

Figure 8. Experimental (solid line) and simulated (dashed line) 2*f* wavelength modulation spectra of 0.01% CH_4 at around 6,046.95 cm^{-1} and a total pressure of 15 kPa diluted by N_2.

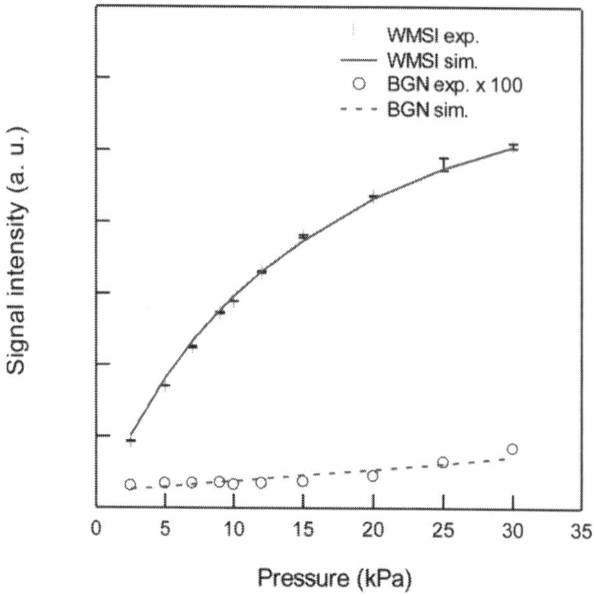

Figure 9. Total pressure dependence of 2*f* wavelength modulation signal intensities (WMSI; •, ———) and background noise (BGN; ○, – – –) for 0.01% CH_4 at 6,046.95 cm^{-1} diluted by N_2. Circles and lines indicate experimental and simulated results, respectively.

5. CONCLUSIONS

We have described the highly sensitive detection of CO_2, N_2O and CH_4 by 2*f* WMS, using tunable near-infrared diode lasers in the near-infrared region. The observed 2*f* WM spectra were analyzed as a function of the total pressure. At the optimum total pressure for the detection of each gas, as determined by experimental and theoretical approaches, the limits of detection in the present system were determined. Detection of atmospheric N_2O (320 ppbv) at the $3\nu_3$ overtone band is not sufficient with our current WMS spectrometer. However, the necessary sensitivity could be delivered by reducing the optical fringe noise and the drifting of laser power using external modulators instead of direct laser modulation.

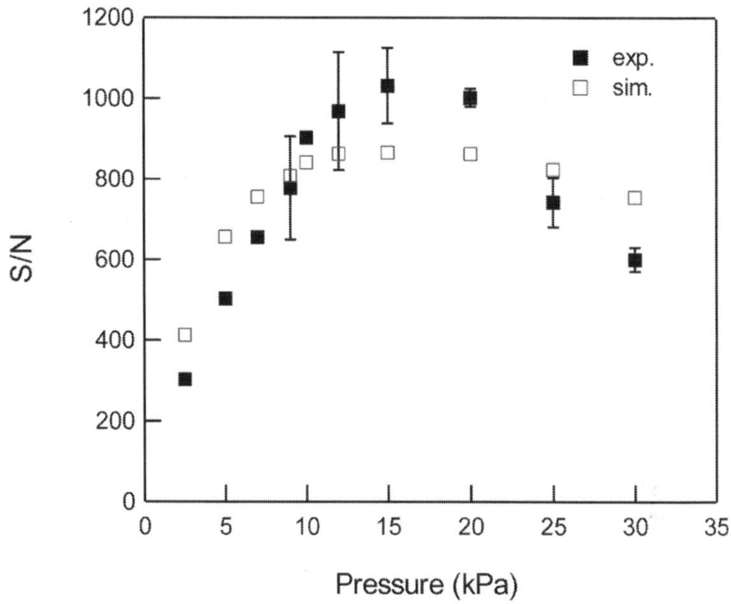

Figure 10. Total pressure dependence of S/N for 0.01% CH_4 at 6,046.97 cm^{-1} diluted by N_2. Closed and open squares indicate experimental and simulated results, respectively.

Table 2. Absorption lines of R(3) of the CH_4 $2\nu_3$ band at around 6,046.97 cm^{-1}.

Line	Position (cm^{-1})	Integrated absorption cross-section (cm^2 $molecule^{-1}$ cm^{-1})
1	6,046.942	8.17×10^{-22}
2	6,046.953	1.00×10^{-21}
3	6,046.965	1.32×10^{-21}

ACKNOWLEDGMENTS

This work was supported by the Industrial Technology Research Grant program from the New Energy and Industrial Technology Development Organization (NEDO) of Japan. The authors acknowledge the helpful suggestions of the referees.

REFERENCES

1. Richter, D.; Fried, A.; Wert, B.P.; Walega, J.G.; Tittel, F.K. Development of a tunable mid-IR difference frequency laser source for highly sensitive airborne trace gas detection. *Appl. Phys. B* **2002**, *75*, 281–288.

2. Gianfrani, L.; Sasso, A.; Tino, G.M. Monitoring of O_2 and NO_2 using tunable diode lasers in the near-infrared region.*Sens. Actuat. B* **1997**, *38*, 283–285.

3. Uehara, K.; Yamamoto, K.; Kikugawa, T.; Yoshida, N. Isotope analysis of environmental substances by a new laser-spectroscopic method utilizing different path lengths. *Sens. Actuat. B* **2001**, *74*, 173–178.

4. Hennig, O.; Strzoda, R.; Magori, E.; Chemisky, E.; Tump, C.; Fleischer, M.; Meixner, H.; Eisele, I. Hand-held unit for simultaneous detection of methane and ethane based on NIR-absorption spectroscopy. *Sens. Actuat. B* **2003**, *95*, 151–156.

5. Katchalski, T.; Soria, S.; Teitelbaum, E.; Friesem, A.A.; Marowsky, G. Two photon fluorescence sensors based on resonant grating waveguide structures. *Sens. Actuat. B* **2005**, *107*, 121–125.

6. Werle, P. A Review of recent advances in semiconductor laser based gas monitors. *Spectrochim. Acta A* **1998**, *54*, 197–236.

7. Linnerud, I.; Kaspersen, P.; Jager, T. Gas monitoring in the process industry using diode laser spectroscopy. *Appl. Phys. B* **1998**, *67*, 297–305.

8. Werle, P.; Mucke, R.; D'Amato, F.; Lancai, T. Near-infrared trace-gas sensors based on room-temperature diode lasers.*Appl. Phys. B* **1998**, *67*, 307–315.

9. Crosson, E.R. A cavity ring-down analyzer for measuring atmospheric levels of methane, carbon dioxide, and water.*Appl. Phys. B* **2008**, *92*, 403–408.

10. Ye, J.; Ma, L.S.; Hall, J.L. Ultrasensitive detections in atomic and molecular physics: demonstration in molecular overtone spectroscopy.*J. Opt. Soc. Am. B* **1998**, *15*, 6–15.

11. Herriott, D.R.; Schulte, H.J. Folded optical delay lines. *Appl. Opt* **1965**, *4*, 883–889.

12. Pavone, F.S.; Inguscio, M. Frequency- and wavelength-modulation spectroscopies: Comparison of experimental methods using an AlGaAs diode laser. *Appl. Phys. B* **1993**, *56*, 118–122.

13. Cooper, D.E.; Warren, R.E. Frequency modulation spectroscopy with lead-salt diode lasers: a comparison of single-tone and two-tone techniques.*Appl. Opt* **1987**, *26*, 3726–3732.

14. Silver, J.A. Frequency-modulation spectroscopy for trace species detection: theory and comparison among experimental methods. *Appl. Opt* **1992**, *31*, 707–717.

15. Supplee, J.M.; Whittaker, E.A.; Lenth, W. Theoretical description of frequency modulation and wavelength modulation spectroscopy.*Appl. Opt* **1994**, *33*, 6294–6302.

16. Wilson, G.V.H. Modulation broadening of NMR and ESR line shapes.*J. Appl. Phys* **1963**, *34*, 3276–3285.

17. LI-COR Bioscience HP. Available on line: http://www.licor.com/env/Products/GasAnalyzers/li7700/7700.jsp (Access on February 2010).

18. Humlicek, J. Optimized computation of the Voigt and complex probability functions. *J. Quant. Spectrosc. Radiat. Transfer* **1982**, *27*, 437–444.

19. Schreier, F. The Voigt and complex error function: A comparison of computational methods. *J. Quant. Spectrosc. Radiat. Transfer* **1992**, *48*, 743–762.

20. Rothman, L.S.; Jacquemart, D.; Barbe, A.; Benner, D.C.; Birk, M.; Brown, L.R.; Carleer, M.R.; Chackerian, C., Jr.; Chance, K.; Coudert, L.H.; Dana, V.; Devi, V.M.; Flaud, J.-M.; Gamache, R.R.; Goldman, A.; Hartmann, J.-M.; Jucks, K.W.; Maki, A.G.; Mandin, J.-Y.; Massie, S.T.; Orphal, J.; Perrin, A.; Rinsland, C.P.; Smith, M.A.H.; Tennyson, J.; Tolchenov, R.N.; Toth, R.A.; Auwera, J.V.; Varanasi, P.; Wagner, G. The HITRAN 2004 molecular spectroscopic database. *J. Quant. Spectrosc. Radiat. Transfer* **2005**, *96*, 139–204.

21. Reid, J.; Labrie, D. Second-harmonic detection with tunable diode lasers-Comparison of experiment and theory. *Appl. Phys. B* **1981**, *26*, 203–210.

22. Parkes, A.M.; Linsley, A.R.; Orr-Ewing, A.J. Absorption cross-sections and pressure broadening of rotational lines in the $3\nu_3$ band of N_2O determined by diode laser cavity ring-down spectroscopy. *Chem. Phys. Lett* 2003, *377*, 439–444.

INDEX

A

Arterial pulsation, 100

C

Centroid method, 100
CMOS image sensor, 100, 103, 104, 105, 107, 108, 111, 123, 124
composite material, 212, 213
Compound Cavity Modes, 226

D

Denim fading, 176
DIAL, 39, 40, 41, 42, 43, 44, 46, 47, 48, 49, 50, 51, 52, 53, 54, 58, 59, 60, 63, 64, 67
Dilute nitride, 84
diode laser, 44, 97, 130, 133, 151, 152, 196, 211, 212, 214, 215, 221, 247, 260, 264, 265, 266

E

Engraving, 176
External Cavity Modes, 226, 244
External Optical Feedback, 225, 226, 242, 243

F

Fabric cutting, 176
filament winding, 211, 212, 214, 223, 224

G

GaInNAs Mobile display, 84
gas sensing, 246
Graphene, 156, 171
ground-based lidar, 40

H

hybrid silicon platform, 1, 2, 4, 7, 9, 15, 16

I

Iterative Traveling-Wave Model, 226, 228

L

Lang-Kobayashi Model, 226
Laser, 1, 18, 20, 39, 43, 44, 66, 67, 68, 96, 97, 99, 100, 101, 116, 125, 126, 129, 133, 151, 175, 176, 177, 178, 180, 181, 183, 246

laser processing, 212, 217
Laser triangulation, 99, 100, 101, 126

M

Mass customization, 176, 187
microwave photonics, 2, 6, 18
mode-locked laser diodes, 2
MSM, 22, 23, 24, 25, 26, 27, 28

N

Nanobelts, 156
Nanorod, 22, 35
Nanowires, 156

O

optical amplification, 70
optical isolators, 2, 3, 18
Optoelectronics, 95, 156

P

Photodetector, 22
photonic integrated circuits, 2
Pulse frequency, 100

Q

quantum dot, 69, 70, 79, 80, 81, 82

S

Schottky junction, 156, 160, 167, 172
Second harmonic generation, 84
Semiconductor laser, 84
Semiconductor Lasers, 226, 243, 244
semiconductor optical amplifiers, 2, 41, 70, 81
silicon photonics, 2, 15, 18
superluminescent diode, 69, 70, 80, 82

T

thermoplastic matrix composites, 212
thermoplastic prepreg, 212, 218
trace gas sensing, 40
two-section structure, 69, 70, 71, 72, 75, 76, 77

V

Visible III-V semiconductors, 84

W

wavelength modulation spectroscopy, 245, 246, 265

Z

ZnO, 21, 22, 23, 24, 25, 26, 27, 28, 29, 30, 31, 33, 34, 35, 36, 37, 157, 163, 165, 167, 172